Introduction to Statistics
Through Resampling
Methods and R

Introduction to Statistics Through Resampling Methods and R

Second Edition

Phillip I. Good

A John Wiley & Sons, Inc., Publication

Published by John Wiley & Sons, Inc., Hoboken, New Jersey.
Published simultaneously in Canada.

For general information on our other products and services or for technical support, please contact our Customer Care Department within the United States at (800) 762-2974, outside the United States at (317) 572-3993 or fax (317) 572-4002.

Wiley also publishes its books in a variety of electronic formats. Some content that appears in print may not be available in electronic formats. For more information about Wiley products, visit our web site at www.wiley.com.

Library of Congress Cataloging-in-Publication Data:

Good, Phillip I.
   Introduction to statistics through resampling methods and R / Phillip I. Good. – Second edition.
       pages cm
     Includes indexes.
     ISBN 978-1-118-42821-4 (pbk.)
1. Resampling (Statistics)   I. Title.
     QA278.8.G63 2013
     519.5'4–dc23

                                            2012031774

10 9 8 7 6 5 4 3 2 1

Contents

10. Reporting Your Findings   **181**

11. Problem Solving   **197**

Preface

Intended for class use or self-study, the second edition of this text aspires as the first to introduce statistical methodology to a wide audience, simply and intuitively, through resampling from the data at hand.

The methodology proceeds from chapter to chapter from the simple to the complex. The stress is always on concepts rather than computations. Similarly, the R code introduced in the opening chapters is simple and straightforward; R's complexities, necessary only if one is programming one's own R functions, are deferred to Chapter 7 and Chapter 8.

The resampling methods—the bootstrap, decision trees, and permutation tests— are easy to learn and easy to apply. They do not require mathematics beyond introductory high school algebra, yet are applicable to an exceptionally broad range of subject areas.

Although introduced in the 1930s, the numerous, albeit straightforward calculations that resampling methods require were beyond the capabilities of the primitive calculators then in use. They were soon displaced by less powerful, less accurate approximations that made use of tables. Today, with a powerful computer on every desktop, resampling methods have resumed their dominant role and table lookup is an anachronism.

Physicians and physicians in training, nurses and nursing students, business persons, business majors, research workers, and students in the biological and social sciences will find a practical and easily grasped guide to descriptive statistics, estimation, testing hypotheses, and model building.

For advanced students in astronomy, biology, dentistry, medicine, psychology, sociology, and public health, this text can provide a first course in statistics and quantitative reasoning.

For mathematics majors, this text will form the first course in statistics to be followed by a second course devoted to distribution theory and asymptotic results.

Hopefully, all readers will find my objectives are the same as theirs: *To use quantitative methods to characterize, review, report on, test, estimate, and classify findings.*

Warning to the autodidact: You can master the material in this text without the aid of an instructor. But you may not be able to grasp even the more elementary concepts without completing the exercises. Whenever and wherever you encounter an exercise in the text, stop your reading and complete the exercise before going further. To simplify the task, R code and data sets may be downloaded by entering ISBN 9781118428214 at booksupport.wiley.com and then cut and pasted into your programs.

I have similar advice for instructors. You can work out the exercises in class and show every student how smart you are, but it is doubtful they will learn anything from your efforts, much less retain the material past exam time. Success in your teaching can be achieved only via the discovery method, that is, by having the students work out the exercises on their own. I let my students know that the final exam will consist solely of exercises from the book. "I may change the numbers or combine several exercises in a single question, but if you can answer all the exercises you will get an A." I do not require students to submit their homework but merely indicate that if they wish to do so, I will read and comment on what they have submitted. When a student indicates he or she has had difficulty with an exercise, emulating Professor Neyman I invite him or her up to the white board and give hints until the problem has been worked out by the student.

Thirty or more exercises included in each chapter plus dozens of thought-provoking questions in Chapter 11 will serve the needs of both classroom and self-study. The discovery method is utilized as often as possible, and the student and conscientious reader forced to think his or her way to a solution rather than being able to copy the answer or apply a formula straight out of the text.

Certain questions lend themselves to in-class discussions in which all students are encouraged to participate. Examples include Exercises 1.11, 2.7, 2.24, 2.32, 3.18, 4.1, 4.11, 6.1, 6.9, 9.7, 9.17, 9.30, and all the problems in Chapter 11.

R may be downloaded without charge for use under Windows, UNIX, or the Macintosh from http://cran.r-project.org. For a one-quarter short course, I take students through Chapter 1, Chapter 2, and Chapter 3. Sections preceded by an asterisk (*) concern specialized topics and may be skipped without loss in comprehension. We complete Chapter 4, Chapter 5, and Chapter 6 in the winter quarter, finishing the year with Chapter 7, Chapter 8, and Chapter 9. Chapter 10 and Chapter 11 on "Reports" and "Problem Solving" convert the text into an invaluable professional resource.

If you find this text an easy read, then your gratitude should go to the late Cliff Lunneborg for his many corrections and clarifications. I am deeply indebted to Rob J. Goedman for his help with the R language, and to Brigid McDermott, Michael L. Richardson, David Warton, Mike Moreau, Lynn Marek, Mikko Mönkkönen, Kim Colyvas, my students at UCLA, and the students in the Introductory Statistics and Resampling Methods courses that I offer online each quarter through the auspices of statcourse.com for their comments and corrections.

PHILLIP I. GOOD
Huntington Beach, CA
drgood@statcourse.com

Chapter 1

Variation

If there were no variation, if every observation were predictable, a mere repetition of what had gone before, there would be no need for statistics.

In this chapter, you'll learn what statistics is all about, variation and its potential sources, and how to use R to display the data you've collected. You'll start to acquire additional vocabulary, including such terms as accuracy and precision, mean and median, and sample and population.

1.1 VARIATION

We find physics extremely satisfying. In high school, we learned the formula $S = VT$, which in symbols relates the distance traveled by an object to its velocity multiplied by the time spent in traveling. If the speedometer says 60 mph, then in half an hour, you are certain to travel exactly 30 mi. Except that during our morning commute, the speed we travel is seldom constant, and the formula not really applicable. Yahoo Maps told us it would take 45 minutes to get to our teaching assignment at UCLA. Alas, it rained and it took us two and a half hours.

Politicians always tell us the best that can happen. If a politician had spelled out the worst-case scenario, would the United States have gone to war in Iraq without first gathering a great deal more information?

In college, we had Boyle's law, $V = KT/P$, with its tidy relationship between the volume V, temperature T and pressure P of a perfect gas. This is just one example of the perfection encountered there. The problem was we could never quite duplicate this (or any other) law in the Freshman Physics' laboratory. Maybe it was the measuring instruments, our lack of familiarity with the equipment, or simple measurement error, but we kept getting different values for the constant K.

By now, we know that variation is the norm. Instead of getting a fixed, reproducible volume V to correspond to a specific temperature T and pressure P, one ends up with a *distribution* of values of V instead as a result of errors in measurement. But we also know that with a large enough representative *sample* (defined later in this chapter), the center and shape of this distribution are reproducible.

Introduction to Statistics Through Resampling Methods and R, Second Edition. Phillip I. Good.
© 2013 John Wiley & Sons, Inc. Published 2013 by John Wiley & Sons, Inc.

Here's more good and bad news: Make astronomical, physical, or chemical measurements and the only variation appears to be due to observational error. Purchase a more expensive measuring device and get more precise measurements and the situation will improve.

But try working with people. Anyone who spends any time in a schoolroom—whether as a parent or as a child, soon becomes aware of the vast differences among individuals. Our most distinct memories are of how large the girls were in the third grade (ever been beat up by a girl?) and the trepidation we felt on the playground whenever teams were chosen (not right field again!). Much later, in our college days, we were to discover there were many individuals capable of devouring larger quantities of alcohol than we could without noticeable effect. And a few, mostly of other nationalities, whom we could drink under the table.

Whether or not you imbibe, we're sure you've had the opportunity to observe the effects of alcohol on others. Some individuals take a single drink and their nose turns red. Others can't seem to take just one drink.

Despite these obvious differences, scheduling for outpatient radiology at many hospitals is done by a computer program that allots exactly 15 minutes to each patient. Well, I've news for them and their computer. Occasionally, the technologists are left twiddling their thumbs. More often the waiting room is overcrowded because of routine exams that weren't routine or where the radiologist wanted additional X-rays. (To say nothing of those patients who show up an hour or so early or a half hour late.)

The majority of effort in *experimental design*, the focus of Chapter 6 of this text, is devoted to finding ways in which this variation from individual to individual won't swamp or mask the variation that results from differences in treatment or approach. It's probably safe to say that what distinguishes statistics from all other branches of applied mathematics is that it is devoted to characterizing and then accounting for *variation* in the observations.

Consider the Following Experiment

You catch three fish. You heft each one and estimate its weight; you weigh each one on a pan scale when you get back to the dock, and you take them to a chemistry laboratory and weigh them there. Your two friends on the boat do exactly the same thing. (All but Mike; the chemistry professor catches him in the lab after hours and calls campus security. This is known as missing data.)

The 26 weights you've recorded ($3 \times 3 \times 3 - 1$ when they nabbed Mike) differ as result of measurement error, observer error, differences among observers, differences among measuring devices, and differences among fish.

1.2   COLLECTING DATA

The best way to observe variation is for you, the reader, to collect some data. But before we make some suggestions, a few words of caution are in order: 80% of the effort in any study goes into data collection and preparation for data collection. Any effort you don't expend initially goes into cleaning up the resulting mess. Or, as my carpenter friends put it, "measure twice; cut once."

We constantly receive letters and emails asking which statistic we would use to rescue a misdirected study. We know of no magic formula, no secret procedure known only to statisticians with a PhD. The operative phrase is GIGO: garbage in, garbage out. So think carefully before you embark on your collection effort. Make a list of possible sources of variation and see if you can eliminate any that are unrelated to the objectives of your study. If midway through, you think of a better method—don't use it.* Any inconsistency in your procedure will only add to the undesired variation.

1.2.1 A Worked-Through Example

Let's get started. Suppose we were to record the time taken by an individual to run around the school track. Before turning the page to see a list of some possible sources of variation, test yourself by writing down a list of all the factors you feel will affect the individual's performance. Obviously, the running time will depend upon the individual's sex, age, weight (for height and age), and race. It also will depend upon the weather, as I can testify from personal experience.

Soccer referees are required to take an annual physical examination that includes a mile and a quarter run. On a cold March day, the last time I took the exam in Michigan, I wore a down parka. Halfway through the first lap, a light snow began to fall that melted as soon as it touched my parka. By the third go around the track, the down was saturated with moisture and I must have been carrying a dozen extra pounds. Needless to say, my running speed varied considerably over the mile and a quarter.

As we shall see in the chapter on analyzing experiments, we can't just add the effects of the various factors, for they often interact. Consider that Kenyan's dominate the long-distance races, while Jamaicans and African-Americans do best in sprints.

The sex of the observer is also important. Guys and stallions run a great deal faster if they think a maiden is watching. The equipment the observer is using is also important: A precision stopwatch or an ordinary wrist watch? (See Table 1.1.)

Before continuing with your reading, follow through on at least one of the following data collection tasks or an equivalent idea of your own as we will be using the data you collect in the very next section:

1. **a.** Measure the height, circumference, and weight of a dozen humans (or dogs, or hamsters, or frogs, or crickets).
   b. Alternately, date some rocks, some fossils, or some found objects.

---

* On the other hand, we strongly recommend you do a thorough review *after* all your data have been collected and analyzed. You can and should learn from experience.

Table 1.1   Sources of Variation in Track Results

Test subject	Observer	Environment	Track
Sex	Sex	Wind	Surface
Age	Measuring device	Rain	Length
Weight/height	Experience	Sun	Condition
Experience		Temperature	
Race			
Current Health			
Desire			

2. Time some tasks. Record the times of 5–10 individuals over three track lengths (say, 50 m, 100 m, and a quarter mile). Since the participants (or trial subjects) are sure to complain they could have done much better if only given the opportunity, record at least two times for each study subject. (Feel free to use frogs, hamsters, or turtles in place of humans as runners to be timed. Or to replaces foot races with knot tying, bandaging, or putting on a uniform.)

3. Take a survey. Include at least three questions and survey at least 10 subjects. All your questions should take the form "Do you prefer A to B? Strongly prefer A, slightly prefer A, indifferent, slightly prefer B, strongly prefer B." For example, "Do you prefer Britney Spears to Jennifer Lopez?" or "Would you prefer spending money on new classrooms rather than guns?"

Exercise 1.1: Collect data as described in one of the preceding examples. Before you begin, write down a complete description of exactly what you intend to measure and how you plan to make your measurements. Make a list of all potential sources of variation. When your study is complete, describe what deviations you had to make from your plan and what additional sources of variation you encountered.

1.3   SUMMARIZING YOUR DATA

Learning how to adequately summarize one's data can be a major challenge. Can it be explained with a single number like the median or mean? The *median* is the middle value of the observations you have taken, so that half the data have a smaller value and half have a greater value. Take the observations 1.2, 2.3, 4.0, 3, and 5.1. The observation 3 is the one in the middle. If we have an even number of observations such as 1.2, 2.3, 3, 3.8, 4.0, and 5.1, then the best one can say is that the median or midpoint is a number (any number) between 3 and 3.8. Now, a question for you: what are the median values of the measurements you made in your first exercise?

Hopefully, you've already collected data as described in the preceding section; otherwise, face it, you are behind. Get out the tape measure and the scales. If you conducted time trials, use those data instead. Treat the observations for each of the three distances separately.

If you conducted a survey, we have a bit of a problem. How does one translate "I would prefer spending money on new classrooms rather than guns" into a number a computer can add and subtract? There is more one way to do this, as we'll discuss in what follows under the heading, "Types of Data." For the moment, assign the number 1 to "Strongly prefer classrooms," the number 2 to "Slightly prefer classrooms," and so on.

1.3.1 Learning to Use R

Calculating the value of a statistic is easy enough when we've only one or two observations, but a major pain when we have 10 or more. As for drawing graphs— one of the best ways to summarize your data—many of us can't even draw a straight line. So do what I do: let the computer do the work.

We're going to need the help of a programming language R that is specially designed for use in computing statistics and creating graphs. You can download that language without charge from the website http://cran.r-project.org/. Be sure to download the kind that is specific to your model of computer and operating system.

As you read through the rest of this text, be sure to have R loaded and running on your computer at the same time, so you can make use of the R commands we provide.

R is an *interpreter*. This means that as we enter the lines of a typical program, we'll learn on a line-by-line basis whether the command we've entered makes sense (to the computer) and be able to correct the line if we've made a typing error.

When we run R, what we see on the screen is an arrowhead

```
>
```

If we type 2 + 3 after and then press the enter key, we see

```
[1]  5.
```

This is because R reports numeric results in the form of a vector. In this example, the first and only element in this vector takes the value 5.

To enter the observations 1.2, 2.3, 4.0, 3 and 5.1, type

```
ourdata = c(1.2,  2.3,  4.0,  3,  5.1)
```

If you've never used a programming language before, let us warn you that R is very inflexible. It won't understand (or, worse, may misinterpret) both of the following:

```
ourdata = c(1.2  2.3  4.0  3  5.1)
ourdata = (1.2,  2.3,  4.0,  3,  5.1)
```

If you did type the line correctly, then typing median (ourdata) afterward will yield the answer 3 after you hit the enter key.

```
ourdata = c(1.2  2.3  4.0  3  5.1)
Error: syntax error
ourdata = c(1.2,  2.3,  4.0,  3,  5.1)
median(ourdata)
[1]  3
```

R Functions

median() is just one of several hundred built-in R functions.

You must use parentheses when you make use of an R function and you must spell the function name correctly.

```
> Median()
Error: could not find function "Median"
> median(Ourdata)
Error in median(Ourdata) : object 'Ourdata' not found
```

The median may tell us where the center of a distribution is, but it provides no information about the variability of our observations, and variation is what statistics is all about. Pictures tell the story best.*

The one-way *strip chart* (Figure 1.1) reveals that the *minimum* of this particular set of data is 0.9 and the *maximum* is 24.8. Each vertical line in this strip chart corresponds to an observation. Darker lines correspond to multiple observations. The *range* over which these observations extend is 24.8—0.9 or about 24.

Figure 1.2 shows a combination *box plot* (top section) and one-way strip chart (lower section). The "box" covers the middle 50% of the sample extending from the 25th to the 75th percentile of the distribution; its length is termed the *interquartile range*. The bar inside the box is located at the median or 50th *percentile* of the sample.

A weakness of this figure is that it's hard to tell exactly what the values of the various percentiles are. A glance at the *box and whiskers plot* (Figure 1.3) made with R suggests the median of the classroom data described in Section 1.5 is about 153 cm, and the interquartile range (the "box") is close to 14 cm. The minimum and maximum are located at the ends of the "whiskers."

To illustrate the use of R to create such graphs, in the next section, we'll use some data I gathered while teaching mathematics and science to sixth graders.

Figure 1.1   Strip chart.

Figure 1.2   Combination *box plot* (top section) and one-way strip chart.

---

* The R code you'll need to create graphs similar to Figures 1.1–1.3 is provided in Section 1.4.1.

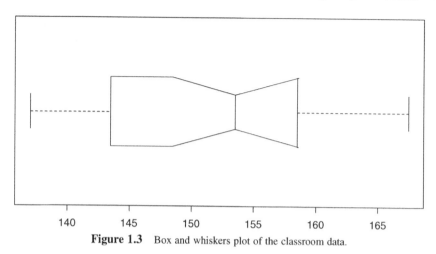

Figure 1.3   Box and whiskers plot of the classroom data.

1.4 REPORTING YOUR RESULTS

Imagine you are in the sixth grade and you have just completed measuring the heights of all your classmates.

Once the pandemonium has subsided, your instructor asks you and your team to prepare a report summarizing your results.

Actually, you have two sets of results. The first set consists of the measurements you made of you and your team members, reported in centimeters, 148.5, 150.0, and 153.0. (Kelly is the shortest incidentally, while you are the tallest.) The instructor asks you to report the minimum, the median, and the maximum height in your group. This part is easy, or at least it's easy once you look the terms up in the glossary of your textbook and discover that *minimum* means smallest, *maximum* means largest, and *median* is the one in the middle. Conscientiously, you write these definitions down—they could be on a test.

In your group, the minimum height is 148.5 cm, the median is 150.0 cm, and the maximum is 153.0 cm.

Your second assignment is more challenging. The results from all your classmates have been written on the blackboard—all 22 of them.

    141, 156.5, 162, 159, 157, 143.5, 154, 158, 140, 142,
    150, 148.5, 138.5, 161, 153, 145, 147, 158.5, 160.5,
    167.5, 155, 137

You copy the figures neatly into your notebook computer. Using R, you store them in classdata using the command,

```
classdata = c(141, 156.5, 162, 159, 157, 143.5, 154,
158, 140, 142, 150, 148.5, 138.5, 161, 153, 145, 147,
158.5, 160.5, 167.5, 155, 137)
```

Next, you brainstorm with your teammates. Nothing. Then John speaks up—he's always interrupting in class. "Shouldn't we put the heights in order from smallest to largest?"

"Of course," says the teacher, "you should always begin by ordering your observations."

```
➤ sort(classdata)
[1] 137.0 138.5 140.0 141.0 142.0 143.5 145.0 147.0
148.5 150.0 153.0 154.0
[13] 155.0 156.5 157.0 158.0 158.5 159.0 160.5 161.0
162.0 167.5
```

In R, when the resulting output takes several lines, the position of the output item in the data set is noted at the beginning of the line. Thus, 137.0 is the first item in the ordered set classdata, and 155.0 is the 13th item.

"I know what the minimum is," you say—come to think of it, you are always blurting out in class, too, "137 millimeters, that's Tony."

"The maximum, 167.5, that's Pedro, he's tall," hollers someone from the back of the room.

As for the median height, the one in the middle is just 153 cm (or is it 154)? What does R say?

```
➤ median(classdata)
```

It is a custom among statisticians, honored by R, to report the median as the value midway between the two middle values, when the number of observations is even.

1.4.1 Picturing Data

The preceding scenario is a real one. The results reported here, especially the pandemonium, were obtained by my sixth-grade homeroom at St. John's Episcopal School in Rancho Santa Marguerite CA. The problem of a metric tape measure was solved by building their own from string and a meter stick.

My students at St. John's weren't through with their assignments. It was important for them to build on and review what they'd learned in the fifth grade, so I had them draw pictures of their data. Not only is drawing a picture fun, but pictures and graphs are an essential first step toward recognizing patterns.

Begin by constructing both a strip chart and a box and whiskers plot of the classroom data using the R commands

```
➤ stripchart(classdata)
```

and

```
➤ boxplot(classdata)
```

All R plot commands have options that can be viewed via the R HELP menu. For example, Figure 1.4 was generated with the command

Figure 1.4   Getting help from R with using R.

> boxplot(classdata, notch=TRUE, horizontal =TRUE)

Generate a strip chart and a box plot for one of the data sets you gathered in your initial assignment. Write down the values of the median, minimum, maximum, 25th and 75th percentiles that you can infer from the box plot. Of course, you could also obtain these same values directly by using the R command, **quantile**(classdata), which yields all the desired statistics.

```
0%        25%       50%       75%       100%
137.000  143.875  153.500  158.375  167.500
```

One word of caution: R (like most statistics software) yields an excessive number of digits. Since we only measured heights to the nearest centimeter, reporting the 25th percentile as 143.875 suggests far more precision in our measurements than what actually exists. Report the value 144 cm instead.

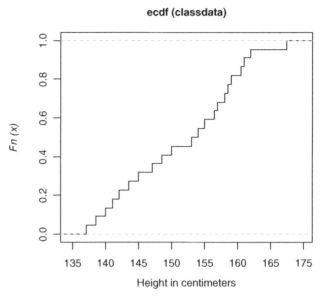

ecdf (classdata)

Figure 1.5 Cumulative distribution of heights of sixth-grade class.

A third way to depict the distribution of our data is via the histogram:

➤ hist(classdata)

To modify a histogram by increasing or decreasing the number of bars that are displayed, we make use of the "breaks" parameter as in

➤ hist(classdata, breaks = 4)

Still another way to display your data is via the *cumulative distribution function* **ecdf**(). To display the cumulative distribution function for the classdata, type

➤ plot(ecdf(classdata), do.points = FALSE, verticals = TRUE, xlab = "Height in Centimeters")

Notice that the *X*-axis of the cumulative distribution function extends from the minimum to the maximum value of your data. The *Y*-axis reveals that the probability that a data value is less than the minimum is 0 (you knew that), and the probability that a data value is less than the maximum is 1. Using a ruler, see what *X*-value or values correspond to 0.5 on the *Y*-scale (Figure 1.5).

Exercise 1.2: What do we call this *X*-value(s)?

Exercise 1.3: Construct histograms and cumulative distribution functions for the data you've collected.

1.4.2 Better Graphics

To make your strip chart look more like the ones shown earlier, you can specify the use of a vertical line as the character to be used in plotting the points:

> ➤ `stripchart(classdata,pch = "|")`

And you can create a graphic along the lines of Figure 1.2, incorporating both a box plot and strip chart, with these two commands

> ➤ `boxplot(classdata,horizontal = TRUE,xlab = "classdata")`
> ➤ `rug(classdata)*`

The first command also adds a label to the *x*-axis, giving the name of the data set, while the second command adds the strip chart to the bottom of the box plot.

1.5 TYPES OF DATA

Statistics such as the minimum, maximum, median, and percentiles make sense only if the data are *ordinal*, that is, if the data can be ordered from smallest to largest. Clearly height, weight, number of voters, and blood pressure are ordinal. So are the answers to survey questions, such as "How do you feel about President Obama?"

Ordinal data can be subdivided into metric and nonmetric data. *Metric* data or measurements like heights and weights can be added and subtracted. We can compute the mean as well as the median of metric data. (Statisticians further subdivide metric data into observations such as time that can be measured on a *continuous* scale and counts such as "buses per hour" that take *discrete* values.)

But what is the average of "he's destroying our country" and "he's no worse than any other politician?" Such preference data are ordinal, in that the data may be ordered, but they are *not* metric.

In order to analyze ordinal data, statisticians often will impose a metric on the data—assigning, for example, weight 1 to "Bush is destroying our country" and weight 5 to "Bush is no worse than any other politician." Such analyses are suspect, for another observer using a different set of weights might get quite a different answer.

The answers to other survey questions are not so readily ordered. For example, "What is your favorite color?" Oops, bad example, as we can associate a metric wavelength with each color. Consider instead the answers to "What is your favorite breed of dog?" or "What country do your grandparents come from?" The answers to these questions fall into nonordered categories. Pie charts and bar charts are used to display such *categorical* data, and contingency tables are used to analyze it. A scatter plot of categorical data would not make sense.

Exercise 1.4: For each of the following, state whether the data are metric and ordinal, only ordinal, categorical, or you can't tell:

- **a.** Temperature
- **b.** Concert tickets
- **c.** Missing data
- **d.** Postal codes.

---

* The rug() command is responsible for the tiny strip chart or rug at the bottom of the chart. Sometimes, it yields a warning message that can usually be ignored.

1.5.1 Depicting Categorical Data

Three of the students in my class were of Asian origin, 18 were of European origin (if many generations back), and one was part American Indian. To depict these categories in the form of a pie chart, I first entered the categorical data:

```
> origin = c(3,18,1)
> pie(origin)
```

The result looks correct, that is, *if* the data are in front of the person viewing the chart. A much more informative diagram is produced by the following R code:

```
> origin = c(3,18,1)
> names(origin) = c("Asian","European","Amerind")
> pie (origin, labels = names(origin))
```

All the graphics commands in R have many similar options; use R's help menu shown in Figure 1.4 to learn exactly what these are.

A pie chart also lends itself to the depiction of ordinal data resulting from surveys. If you performed a survey as your data collection project, make a pie chart of your results, now.

1.6 DISPLAYING MULTIPLE VARIABLES

I'd read but didn't quite believe that one's arm span is almost exactly the same as one's height. To test this hypothesis, I had my sixth graders get out their tape measures a second time. They were to rule off the distance from the fingertips of the left hand to the fingertips of the right while the student they were measuring stood with arms outstretched like a big bird. After the assistant principal had come and gone (something about how the class was a little noisy, and though we were obviously having a good time, could we just be a little quieter), they recorded their results in the form of a two-dimensional scatter plot.

They had to reenter their height data (it had been sorted, remember), and then enter their armspan data:

```
> classdata = c(141, 156.5, 162, 159, 157, 143.5,
154, 158, 140, 142, 150, 148.5, 138.5, 161, 153, 145,
147, 158.5, 160.5, 167.5, 155, 137)
> armspan = c(141, 156.5, 162, 159, 158,
143.5, 155.5, 160, 140, 142.5, 148, 148.5, 139,
160, 152.5, 142, 146.5, 159.5, 160.5, 164, 157,
137.5)
```

This is trickier than it looks, because unless the data are entered in exactly the same order by student in each data set, the results are meaningless. (We told you that 90% of the problem is in collecting the data and entering the data in the computer for analysis. In another text of mine, *A Manager's Guide to the Design and Conduct of Clinical Trials*, I recommend eliminating paper forms completely and entering all data directly into the computer.*) Once the two data sets have been read in, creating a scatterplot is easy:

```
➤ height = classdata
➤ plot(height, armspan)
```

Notice that we've renamed the vector we called classdata to reveal its true nature as a vector of heights.

Such plots and charts have several purposes. One is to summarize the data. Another is to compare different samples or different populations (girls versus boys, my class versus your class). For example, we can enter gender data for the students, being careful to enter the gender codes in the same order in which the students' heights and arm spans already have been entered:

```
➤ sex = c("b",rep("g",7),"b",rep("g",6),rep("b",7))
```

Note that we've introduced a new R function, **rep**(), in this exercise to spare us having to type out the same value several times. The first student on our list is a boy; the next seven are girls, then another boy, six girls, and finally seven boys. R requires that we specify non-numeric or *character data* by surrounding the elements with quote signs. We then can use these gender data to generate side-by-side box plots of height for the boys and girls.

```
➤ sexf = factor(sex)
➤ plot(sexf,height)
```

The R-function **factor** () tells the computer to treat gender as a categorical variable, one that in this case takes two values "b" and "g." The **plot**() function will not work until character data have been converted to factors.

The primary value of charts and graphs is as aids to critical thinking. The figures in this specific example may make you start wondering about the uneven way adolescents go about their growth. The exciting thing, whether you are a parent or a middle-school teacher, is to observe how adolescents get more heterogeneous, more individual with each passing year.

Exercise 1.5: Use the preceding R code to display and examine the indicated charts for my classroom data.

Exercise 1.6: Modify the preceding R code to obtain side-by-side box plots for the data you've collected.

1.6.1 Entering Multiple Variables

We've noted several times that the preceding results make sense *only* if the data are entered in the same order by student for each variable that you are interested in. R provides a simple way to achieve this goal. Begin by writing,

---

* We'll discuss how to read computer data files using R later in this chapter.

Figure 1.6 The R edit() screen.

```
> student = 1:length(height)
> classdata = data.frame(student, height)
> classdata = edit(classdata)
```

The upper left corner of R's screen will then resemble Figure 1.6. While still in edit mode, we add data for arm length and sex until the screen resembles Figure 1.7.

To rename variables, simply click on the existing name. Enter the variable's name, then note whether the values you enter for that variable are to be treated as numbers or characters.

1.6.2 From Observations to Questions

You may want to formulate your theories and suspicions in the form of questions: Are girls in the sixth-grade taller on the average than sixth-grade boys (not just those in my sixth-grade class, but in all sixth-grade classes)? Are they more homogenous, that is, less variable, in terms of height? What is the average height of a sixth grader? How reliable is this estimate? Can height be used to predict arm span in sixth grade? Can it be used to predict the arm spans of students of any age?

You'll find straightforward techniques in subsequent chapters for answering these and other questions. First, we suspect, you'd like the answer to one really big question: Is statistics really much more difficult than the sixth-grade exercise we just completed? No, this is about as complicated as it gets.

Figure 1.7   The R edit() screen after entering additional observations.

1.7 MEASURES OF LOCATION

Far too often, we find ourselves put on the spot, forced to come up with a one-word description of our results when several pages, or, better still, several charts would do. "Take all the time you like," coming from a boss, usually means, "Tell me in ten words or less."

If you were asked to use a single number to describe data you've collected, what number would you use? One answer is "the one in the middle," the *median* that we defined earlier in this chapter. The median is the best statistic to use when the data are *skewed*, that is, when there are unequal numbers of small and large values. Examples of skewed data include both house prices and incomes.

In most other cases, we recommend using the *arithmetic mean* rather than the median.* To calculate the mean of a sample of observations by hand, one adds up the values of the observations, then divides by the number of observations in the sample. If we observe 3.1, 4.5, and 4.4, the arithmetic mean would be 12/3 = 4. In symbols, we write the mean of a sample of n observations, X_i with $i = 1, 2, \ldots,$ n as

$$(X_1 + X_2 + \cdots + X_n)/n = \frac{1}{n}\sum_{i=1}^{n} X_i = \bar{X}.^\dagger$$

---

* Our reason for this recommendation will be made clear in Chapter 5.
$\dagger$ The Greek letter Σ is pronounced sigma.

Is adding a set of numbers and then dividing by the number in the set too much work? To find the mean height of the students in Dr. Good's classroom, use R and enter

> ➤ mean(height).

A playground seesaw (or teeter-totter) is symmetric in the absence of kids. Its midpoint or median corresponds to its center of gravity or its mean. If you put a heavy kid at one end and two light kids at the other so that the seesaw balances, the mean will still be at the pivot point, but the median is located at the second kid.

Another population parameter of interest is the most frequent observation or *mode*. In the sample 2, 2, 3, 4 and 5, the mode is 2. Often the mode is the same as the median or close to it. Sometimes it's quite different and sometimes, particularly when there is a mixture of populations, there may be several modes.

Consider the data on heights collected in my sixth-grade classroom. The mode is at 157.5 cm. But aren't there really two modes, one corresponding to the boys, the other to the girls in the class? As you can see on typing the command

> ➤ hist(classdata, xlab = "Heights of Students in Dr. Good's Class (cms)")

a histogram of the heights provides evidence of two modes. When we don't know in advance how many subpopulations there are, modes serve a second purpose: to help establish the number of subpopulations.

Exercise 1.7: Compare the mean, median, and mode of the data you've collected (Figure 1.8).

Histogram of classdata

Heights of students in Dr. Good's class (cms)

Figure 1.8 Histogram of heights of sixth-grade students.

Exercise 1.8: A histogram can be of value in locating the modes when there are 20 to several hundred observations, because it groups the data. Use R to draw histograms for the data you've collected.

1.7.1 Which Measure of Location?

The arithmetic mean, the median, and the mode are examples of sample *statistics*. Statistics serve three purposes:

1. Summarizing data
2. Estimating population parameters
3. Aids to decision making.

Our choice of one statistic rather than another depends on the use(s) to which it is to be put.

The Center of a Population

Median. The value in the middle; the halfway point; that value which has equal numbers of larger and smaller elements around it.

Arithmetic Mean or Arithmetic Average. The sum of all the elements divided by their number or, equivalently, that value such that the sum of the deviations of all the elements from it is zero.

Mode. The most frequent value. If a population consists of several subpopulations, there may be several modes.

For summarizing data, graphs—box plots, strip plots, cumulative distribution functions, and histograms are essential. If you're not going to use a histogram, then for samples of 20 or more, be sure to report the number of modes.

We always recommend using the median in two instances:

1. If the data are ordinal but not metric.
2. When the distribution of values is highly skewed with a few very large or very small values.

Two good examples of the latter are incomes and house prices. A recent *LA Times* featured a great house in Beverly Hills at US$80 million. A house like that has a large effect on the mean price of homes in an area. The median house price is far more representative than the mean, even in Beverly Hills.

The weakness of the arithmetic mean is that it is too easily biased by extreme values. If we eliminate Pedro from our sample of sixth graders—he's exceptionally tall for his age at 5'7"—the mean would change from 151.6 to 3167/21 = 150.8 cm. The median would change to a much lesser degree, shifting from 153.5 to 153 cm.

Because the median is not as readily biased by extreme values, we say that the median is more *robust* than the mean.

*1.7.2 The Geometric Mean

The *geometric mean* is the appropriate measure of location when we are expressing changes in percentages, rather than absolute values. The geometric mean's most common use is in describing bacterial and viral populations.

Here is another example: If in successive months the cost of living was 110, 105, 110, 115, 118, 120, 115% of the value in the base month, set

> ➤ ourdata = c(1.1,1.05,1.1,1.15,1.18,1.2,1.15)

The geometric mean is given by the following R expression:

> ➤ exp(mean(log(ourdata))

Just be sure when you write expressions as complicated as the expression in the line above, that the number of right parentheses (matches the number of left parentheses).

For estimation: In deciding which *sample statistic* to use in estimating the corresponding *population parameter*, we must distinguish between precision and accuracy. Let us suppose Robin Hood and the Sheriff of Nottingham engage in an archery contest. Each is to launch three arrows at a target 50 m (half a soccer pitch) away. The Sheriff launches first and his three arrows land one atop the other in a dazzling display of shooting *precision*. Unfortunately all three arrows penetrate and fatally wound a cow grazing peacefully in the grass nearby. The Sheriff's *accuracy* leaves much to be desired.

1.7.3 Estimating Precision

We can show mathematically that for very large samples, the sample median and the population median will almost coincide. The same is true for large samples and the mean. Alas, "large" may mean larger than we can afford to examine. With small samples, the accuracy of an estimator is always suspect.

With most of the samples we encounter in practice, we can expect the value of the sample median and virtually any other estimator to vary from sample to sample. One way to find out for small samples how *precise* a method of estimation is would be to take a second sample the same size as the first and see how the estimator varies between the two. Then a third, and fourth, . . . , say, 20 samples. *But a large sample will* always *yield more precise results than a small one.* So, if we'd been able to afford it, the sensible thing would have been to take 20 times as large a sample to begin with.*

Still, there is an alternative. We can treat our sample as if it were the original population and take a series of *bootstrap samples* from it. The variation in the value

---

* Of course, there is a point at which each additional observation will cost more than it yields in information. The bootstrap described here will also help us to find the "optimal" sample size.

of the estimator from bootstrap sample to bootstrap sample will be a measure of the variation to be expected in the estimator had we been able to afford to take a series of samples from the population itself. The larger the size of the original sample, the closer it will be in composition to the population from which it was drawn, and the more accurate this measure of precision will be.

1.7.4 Estimating with the Bootstrap

Let's see how this process, called bootstrapping, would work with a specific set of data. Once again, here are the heights of the 22 students in Dr. Good's sixth grade class, measured in centimeters and ordered from shortest to tallest:

```
137.0 138.5 140.0 141.0 142.0 143.5 145.0 147.0 148.5
150.0 153.0 154.0 155.0 156.5 157.0 158.0 158.5 159.0
160.5 161.0 162.0 167.5
```

Let's assume we record each student's height on an index card, 22 index cards in all. We put the cards in a big hat, shake them up, pull one out, and make a note of the height recorded on it. We *return the card to the hat* and repeat the procedure for a total of 22 times until we have a second sample, the same size as the original. Note that we may draw Jane's card several times as a result of using this method of *sampling with replacement*.

Our first bootstrap sample, arranged in increasing order of magnitude for ease in reading, might look like this:

```
138.5 138.5 140.0 141.0 141.0 143.5 145.0 147.0 148.5
150.0 153.0 154.0 155.0 156.5 157.0 158.5 159.0 159.0
159.0 160.5 161.0 162.
```

Several of the values have been repeated; not surprising, as we are sampling with replacement, treating the original sample as a stand-in for the much larger population from which the original sample was drawn. The minimum of this bootstrap sample is 138.5, higher than that of the original sample; the maximum at 162.0 is less than the original, while the median remains unchanged at 153.5.

```
137.0 138.5 138.5 141.0 141.0 142.0 143.5 145.0 145.0
147.0 148.5 148.5 150.0 150.0 153.0 155.0 158.0 158.5
160.5 160.5 161.0 167.5
```

In this second bootstrap sample, again we find repeated values—quick, what are they? This time the minimum, maximum, and median are 137.0, 167.5, and 148.5, respectively.

Two bootstrap samples cannot tell us very much. But suppose we were to take 50 or a hundred such samples. Figure 1.9 shows a one-way strip plot of the medians of 50 bootstrap samples taken from the classroom data. These values provide an

Figure 1.9   One-way strip plot of the medians of 50 bootstrap samples taken from the classroom data.

insight into what might have been had we sampled repeatedly from the original population.

Quick question: What is that population? Does it consist of all classes at the school where I was teaching? All sixth-grade classes in the district? All sixth-grade classes in the state? The school was Episcopalian, and the population from which its pupils were drawn is typical of middle- to upper-class Southern California, so perhaps the population was all sixth-grade classes in Episcopalian schools in middle- to upper-class districts of Southern California.

> **Exercise 1.9:** Our original question, you'll recall, is which is the least variable (or, equivalently, the most precise) estimate: mean or median? To answer this question, at least for several samples, let us apply the bootstrap, first to our classroom data and then to the data we collected in Exercise 1.1. You'll need the following R listing:

```
#Comments to R code all start with the pound sign #
#This program selects 100 bootstrap samples from your
#data and then produces a boxplot of the results.
#first, we give a name, urdata, to the observations
#in our original sample
➤  urdata = c( , . . .)
#Record group size
➤  n = length(urdata)
#set number of bootstrap samples
➤  N = 100
#create a vector of length N in which to store the
#results
➤  stat = numeric(N)
#Set up a loop using the for() function to generate a
#series of bootstrap samples
#All the commands between the brackets { } will be
#repeated N times in this example; the first time i =
#1, the next time i = i + 1.
➤  for (i in 1:N){
#bootstrap sample counterparts to observed samples
#are denoted with "B"
+  urdataB = sample (urdata, n, replace = T)
+  stat[i] = mean(urdataB)
+  }
➤  boxplot (stat)
```

1.8   SAMPLES AND POPULATIONS

If it weren't for person-to-person variation, it really would be easy to find out what brand of breakfast cereal people prefer or which movie star they want as their leader. Interrogate the first person you encounter on the street and all will be revealed. As things stand, either we must pay for and take a total census of everyone's view (the cost of the 2003 recall election in California pushed an already near-bankrupt state one step closer to the edge) or take a sample and learn how to extrapolate from that sample to the entire population.

In each of the data collection examples in Section 1.2, our observations were limited to a sample from a population. We measured the height, circumference, and weight of a dozen humans (or dogs, or hamsters, or frogs, or crickets), but not all humans or dogs or hamsters. We timed some individuals (or frogs or turtles) in races but not all. We interviewed some fellow students but not all.

If we had interviewed a different set of students, would we have gotten the same results? Probably not. Would the means, medians, interquartile ranges, and so forth have been similar for the two sets of students? Maybe, *if* the two samples had been large enough and similar to each other in composition.

If we interviewed a sample of women and a sample of men regarding their views on women's right to choose, would we get similar answers? Probably not, as these samples were drawn from completely different populations (different, i.e., with regard to their views on women's right to choose.) If we want to know how the citizenry as a whole feels about an issue, we need to be sure to interview both men and women.

In every statistical study, two questions immediately arise:

1. How large should my sample be?
2. How can I be sure this sample is representative of the population in which my interest lies?

By the end of Chapter 6, we'll have enough statistical knowledge to address the first question, but we can start now to discuss the second.

After I deposited my ballot in a recent election, I walked up to the interviewer from the *LA Times* who was taking an exit poll and offered to tell her how I'd voted. "Sorry," she said, "I can only interview every ninth person."

What kind of a survey wouldn't want my views? Obviously, a survey that wanted to ensure that shy people were as well represented as boisterous ones and that a small group of activists couldn't bias the results.*

One sample we would all insist be representative is the jury.[†] The Federal Jury Selection and Service Act of 1968 as revised[‡] states that citizens cannot be disqualified from jury duty "on account of race, color, religion, sex, national origin or economic status."[§] The California Code of Civil Procedure, section 197, tells us *how* to get a representative sample. First, you must be sure your sample is taken from the appropriate population. In California's case, the "list of registered voters and the Department of Motor Vehicles list of licensed drivers and identification card holders . . . shall be considered inclusive of a representative cross section of the population." The Code goes on to describe how a table of random numbers or a computer could be used to make the actual selection. The bottom line is that to obtain a random, representative sample:

---

* To see how surveys could be biased deliberately, you might enjoy reading Grisham's *The Chamber*.
[†] Unless of course, we are the ones on trial.
[‡] 28 U.S.C.A. x1861 et. seq (1993).
[§] See 28 U.S.C.A. x1862 (1993).

- Each individual (or item) in the eligible population must have an equal probability of being selected.
- No individual (item) or class of individuals may be discriminated against.

There's good news and bad news. The bad news is that any individual sample may not be representative. You can flip a coin six times and every so often it will come up heads six times in a row.* A jury may consist entirely of white males. The good news is that as we draw larger and larger samples, the samples will resemble more and more closely the population from which they are drawn.

Exercise 1.10: For each of the three data collection examples of Section 1.2, describe the populations you would hope to extend your conclusions to and how you would go about ensuring that your samples were representative in each instance.

1.8.1   Drawing a Random Sample

Recently, one of our clients asked for help with an audit. Some errors had been discovered in an invoice they'd submitted to the government for reimbursement. Since this client, an HMO, made hundreds of such submissions each month, they wanted to know how prevalent such errors were. Could we help them select a sample for analysis?

We could, but we needed to ask the client some questions first. We had to determine what the population was from which the sample would be taken and what constituted a *sampling unit*.

Were they interested in all submissions or just some of them? The client told us that some submissions went to state agencies and some to Federal, but for audit purposes, their sole interest was in certain Federal submissions, specifically in submissions for reimbursement for a certain type of equipment. Here too, a distinction needed to be made between custom equipment (with respect to which there was virtually never an error) and more common off-the-shelf supplies. At this point in the investigation, our client breathed a sigh of relief. We'd earned our fee, it appeared, merely by observing that instead of 10,000 plus potentially erroneous claims, the entire population of interest consisted of only 900 or so items.

(When you read earlier that 90% of the effort in statistics was in designing the experiment and collecting the data, we meant exactly that.)

Our client's staff, like that of most businesses, was used to working with an electronic spreadsheet. "Can you get us a list of all the files in spreadsheet form?" we asked.

They could and did. The first column of the spreadsheet held each claim's ID. The second held the date. We used the spreadsheet's sort function to sort all the

---

* Once you've completed the material in the next chapter, you'll be able to determine the probability of six heads in a row.

claims by date, and then deleted all those that fell outside the date range of interest. Next, a new column was inserted, and in the top cell (just below the label row) of the new column, we put the Excel command **rand**(). We copied this command all the way down the column.

A series of numbers between 0 and 1 was displayed down the column. To lock these numbers in place, we went to the Tools menu, clicked on "options," and then on the calculation tab. Next, we made sure that Calculation was set to manual and there was no check mark opposite "recalculate before save."

Now, we resorted the data based on the results of this column. Beforehand, we'd decided there would be exactly 35 claims in the sample, so we simply cut and pasted the top 35 items.

With the help of R, we can duplicate this entire process with a single command:

```
> randsamp = sample(Data).
```

*1.8.2 Using Data That Are Already in Spreadsheet Form

Use Excel's "Save As" command to save your spreadsheet in csv format. Suppose you have saved it in c:/mywork/Data.dat. Then use the following command to bring the data into R:

```
> Data = read.table("c:/mywork/Data.dat", sep=",")
```

1.8.3 Ensuring the Sample Is Representative

Exercise 1.11: We've already noted that a random sample might not be representative. By chance alone, our sample might include men only, or African-Americans but no Asians, or no smokers. How would you go about ensuring that a random sample is representative?

1.9 SUMMARY AND REVIEW

In this chapter, you learned R's syntax and a number of R commands with which to

- Manipulate data and create vectors of observations (c, edit, sort, numeric, factor)
- Perform mathematical functions (exp,log)
- Compute statistics (mean, median, quantile)
- Create graphs (boxplot, hist, pie, plot, plotCDF, stripchart, rug).

You learned that these commands have parameters such as xlab and ylab that allow you to create more attractive graphs

- Control program flow (for)
- Select random samples (sample)
- Read data from tables (read.table).

The best way to summarize and review the statistical material we've covered so far is with the aid of three additional exercises.

Exercise 1.12: Make a list of all the *italicized* terms in this chapter. Provide a definition for each one along with an example.

Exercise 1.13: The following data on the relationship of performance on the LSATs to GPA are drawn from a population of 82 law schools. We'll look at this data again in Chapters 3 and 4.

```
LSAT = c(576,635,558,578,666,580,555,661,651,605,653,5
75,545,572,594)
GPA = c(3.39,3.3,2.81,3.03,3.44,3.07,3,3.43,3.36,3.13,
3.12,2.74,2.76,2.88,2.96)
```

Make box plots and histograms for both the LSAT score and GPA. Tabulate the mean, median, interquartile range, and the 95th and 5th percentiles for both variables.

Exercise 1.14: I have a theory that literally all aspects of our behavior are determined by our birth order (oldest/only, middle, and youngest) including clothing, choice of occupation, and sexual behavior. How would you go about collecting data to prove or disprove some aspect of this theory?

Chapter 2

Probability

In this chapter, you'll learn the rules of probability and apply them to games of chance, jury selection, surveys, and blood types. You'll use R to generate simulated random data and learn how to create your own R functions.

2.1 PROBABILITY

Because of the variation that is inherent in the processes we study, we are forced to speak in probabilistic terms rather than absolutes. We talk about the *probability* that a sixth-grader is exactly 150 cm tall, or, more often, the probability that his height will lie between two values, such as 150 and 155 cm. The events we study may happen a large proportion of the time, or "almost always," but seldom "always" or "never."

Rather arbitrarily, and some time ago, it was decided that probabilities would be assigned a value between 0 and 1, that events that were certain to occur would be assigned probability 1, and that events that would "never" occur would be given probability zero. This makes sense if we interpret probabilities in terms of frequencies; for example, that the probability that a coin turns up heads is the ratio of the number of heads in a very large number of coin tosses to the total number of coin tosses.

When talking about a set of *equally likely* events, such as the probability that a fair coin will come up heads, or an unweighted die will display a "6," this limitation makes a great deal of sense. A coin has two sides; we say the probability it comes up heads is a half and the probability of tails is a half also. $1/2 + 1/2 = 1$, the probability that a coin comes up something.* Similarly, the probability that a six-sided die displays a "6" is $1/6$. The probability it does *not* display a 6 is $1 - 1/6 = 5/6$.

For every dollar you bet, Roulette wheels pay off $36 if you win. This certainly seems fair, until you notice that not only does the wheel have slots for the numbers 1 through 36, but there is a slot for 0, and sometimes for double 0, and for triple

---

* I had a professor at Berkeley who wrote a great many scholarly articles on the subject of "coins that stand on edge," but then that is what professors at Berkeley do.

Introduction to Statistics Through Resampling Methods and R, Second Edition. Phillip I. Good.
© 2013 John Wiley & Sons, Inc. Published 2013 by John Wiley & Sons, Inc.

000 as well. The real probabilities of winning and losing in this latter case are, respectively, 1 chance in 39 and 38/39. In the long run, you lose one dollar 38 times as often as you win $36. Even when you win, the casino pockets your dollar, so that in the long run, the casino pockets $3 for every $39 that is bet. (And from whose pockets does that money come?)

Ah, but you have a clever strategy called a *martingale*. Every time you lose, you simply double your bet. So if you lose a dollar the first time, you lose two dollars the next. Hmm. Since the casino always has more money than you do, you still end up broke. Tell me again why this is a clever strategy.

Exercise 2.1: List the possible ways in which the following can occur:

    a. A specific person, call him Bill, is born on a Wednesday.
    b. Bill and Alice are born on the same day of the week.
    c. Bill and Alice are born on different days of the week.
    d. Bill and Alice play a round of a game called "paper, scissor, stone" and simultaneously display either an open hand, two fingers, or a closed fist.

Exercise 2.2: Match the probabilities with their descriptions. A description may match more than one probability.

a.	−1	**1.**	infrequent
b.	0	**2.**	virtually impossible
c.	0.10	**3.**	certain to happen
d.	0.25 in	**4.**	typographical error
e.	0.50	**5.**	more likely than not
f.	0.80	**6.**	certain
g.	1.0	**7.**	highly unlikely
h.	1.5	**8.**	even odds
		9.	highly likely

To determine whether a gambling strategy or a statistic is optimal, we need to know a few of the laws of probability. These laws show us how to determine the probabilities of combinations of events. For example, if the probability that an event A will occur is $P\{A\}$, then the probability that A won't occur $P\{notA\} = 1 - P\{A\}$. This makes sense since either the event A occurs or it doesn't, and thus $P\{A\} + P\{notA\} = 1$.

We'll also be concerned with the probability that both A and B occur, $P\{A$ and $B\}$, or with the probability that either A occurs or B occurs or both do, $P\{A$ or $B\}$. If two events A and B are *mutually exclusive*, that is, if when one occurs, the other cannot possibly occur, then the probability that A **or** B will occur, $P\{A$ or $B\}$ is the sum of their separate probabilities. (Quick, what is the probability that both A **and** B occur.)* The probability that a six-sided die will show an odd number is thus 3/6 or 1/2. The probability that a six-sided die will *not* show an even number is equal to the probability that a six-sided die will show an odd number.

---

* If A and B are *mutually exclusive*, $P(A$ and $B) = P(A) \times P(B)$. See Section 2.2 for the proof of this statement.

Just because there are exactly two possibilities does not mean that they are equally likely. Otherwise, the United States would never be able to elect a President. Later in this chapter, we'll consider the binomial distribution, which results when we repeat a sequence of trials, each of which can result in exactly one of two possibilities.

2.1.1 Events and Outcomes

An *outcome* is something we can observe. For example, "the coin lands heads" or "an odd number appears on the die." Outcomes are made up of *events* that may or may not be completely observable. The referee tosses the coin into the air; it flips over three times before he catches it and places it face upward on his opposite wrist. "Heads," and Manchester United gets the call. But the coin might also have come up heads had the coin been tossed higher in the air so that it spun three and a half or four times before being caught. A literal infinity of events makes up the single observed outcome, "Heads."

The outcome "an odd number appears on the six-sided die," is composed of three outcomes, 1, 3, and 5, each of which can be the result of any of an infinity of events. By definition, events are mutually exclusive. Outcomes may or may not be mutually exclusive, depending on how we aggregate events.

2.1.2 Venn Diagrams

An excellent way to gain insight into the distinction between events and outcomes and the laws of probability is via the Venn diagram.* Figure 2.1 pictures two

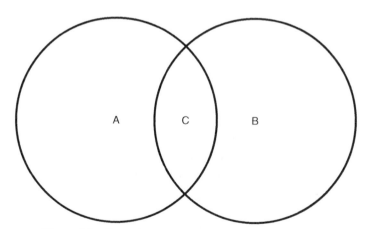

Figure 2.1   Venn Diagram depicting two overlapping outcomes.

---

* Curiously, not a single Venn diagram is to be found in John Venn's text, *The Logic of Chance,* published by Macmillan and Co, London, 1866, with a 3rd edition in 1888.

overlapping outcomes, A and B. For example, the set A might consist of all those who respond to a survey that they are non-smokers, while B corresponds to the set of all respondents that have lung cancer.

Every point in the figure corresponds to an event. The events within the circle A all lead to the outcome A. Note that many of the events or points in the diagram lie outside both circles. These events correspond to the outcome, "neither A nor B" or, in our example, "an individual who does smoke and does not have lung cancer."

The circles overlap; thus outcomes A and B are not mutually exclusive. Indeed, any point in the region of overlap between the two, marked C, leads to the outcome "A and B." What can we say about individuals who lie in region C?

Exercise 2.3: Construct a Venn diagram corresponding to the possible outcomes of throwing a six-sided die. (I find it easier to use squares than circles to represent the outcomes, but the choice is up to you.) Does every event belong to one of the outcomes? Can an event belong to more than of these outcomes? Now, shade the area that contains the outcome, "the number face up on the die is odd." Use a different shading to outline the outcome, "the number on the die is greater than 3."

Exercise 2.4: Are the two outcomes "the number face up on the die is odd" and "the number on the die is greater than 3" mutually exclusive?

You'll find many excellent Venn diagrams illustrating probability concepts at http://stat-www.berkeley.edu/users/stark/Java/Html/Venn.htm

---

In the Long Run: Some Misconceptions

When events occur as a result of chance alone, anything can happen and usually will. You roll craps seven times in a row, or you flip a coin 10 times and 10 times it comes up heads. Both these events are unlikely, but they are not impossible. Before reading the balance of this section, test yourself by seeing if you can answer the following:

You've been studying a certain roulette wheel that is divided into 38 sections for over 4 hours now and not once during those 4 hours of continuous play has the ball fallen into the number 6 slot. Which of the following do you feel is more likely?

1. Number 6 is bound to come up soon.

2. The wheel is fixed so that number 6 will never come up.

3. The odds are exactly what they've always been and in the next four hours number 6 will probably come up about 1/38th of the time.

If you answered (2) or (3) you're on the right track. If you answered (1), think about the following equivalent question:

You've been studying a series of patients treated with a new experimental drug all of whom died in excruciating agony despite the treatment. Do you conclude the drug is bound to cure somebody sooner or later and take it yourself when you come down with the symptoms? Or do you decide to abandon this drug and look for an alternative?

2.2 BINOMIAL TRIALS

Many of our observations take a yes/no or dichotomous form: "My headache did/ didn't get better." "Chicago beat/was beaten by Los Angeles." "The respondent said he would/wouldn't vote for Ron Paul." The simplest example of a *binomial trial* is that of a coin flip: heads I win, tails you lose.

   If the coin is fair, that is, if the only difference between the two mutually exclusive outcomes lies in their names, then the probability of throwing a head is 1/2, and probability of throwing a tail is also 1/2. (That's what I like about my bet, either way I win.)

   By definition, the probability that something will happen is 1, the probability that nothing will occur is 0. All other probabilities are somewhere in between.*

   What about the probability of throwing heads twice in a row? Ten times in a row? If the coin is fair and the throws independent of one another, the answers are easy: 1/4th and 1/1024th or $(1/2)^{10}$.

   These answers are based on our belief that when the only differences among several possible *mutually exclusive* outcomes are their labels, "heads" and "tails," for example, the various outcomes will be *equally likely*. If we flip two fair coins or one fair coin twice in a row, there are four possible outcomes HH, HT, TH, and TT. Each outcome has equal probability of occurring. The probability of observing the one outcome in which we are interested is 1 in 4 or 1/4th. Flip the coin 10 times and there are 2^{10} or a thousand possible outcomes; one such outcome might be described as HTTTTTTTTH.

   Unscrupulous gamblers have weighted coins so that heads comes up more often than tails. In such a case, there is a real difference between the two sides of the coin and the probabilities will be different from those described above. Suppose as a result of weighting the coin, the probability of getting a head is now p, where $0 \le p \le 1$, and the complementary probability of getting a tail (or not getting a head) is $1 - p$, because $p + (1 - p) = 1$. Again, we ask the question, what is the probability of getting two heads in a row? The answer is pp or p^2. Here is why:

   To get two heads in a row, we must throw a head on the first toss, which we expect to do in a proportion p of attempts. Of this proportion, only a further fraction p of two successive tosses also end with a head, that is, only pp trials result in HH. Similarly, the probability of throwing ten heads in a row is p^{10}.

   By the same line of reasoning, we can show the probability of throwing nine heads in a row followed by a tail when we use the same weighted coin each time is $p^9(1 - p)$. What is the probability of throwing 9 heads in 10 trials? Is it also $p^9(1 - p)$? No, for the outcome "nine heads out of ten" includes the case where the first trial is a tail and all the rest are heads, the second trial is a tail and all the rest are heads, the third trial is, and so forth. Ten different ways in all. These different

---

* If you want to be precise, the probability of throwing a head is probably only 0.49999, and the probability of a tail is also only 0.49999. The leftover probability of 0.00002 is the probability of all the other outcomes—the coin stands on edge, a sea gull drops down out of the sky and takes off with it, and so forth.

ways are *mutually exclusive*, that is, if one of these events occurs, the others are excluded. The probability of the overall event is the sum of the individual probabilities or $10\,p^9(1-p)$.

Exercise 2.5: What is the probability that if you flipped a fair coin you would get heads five times in a row?

Exercise 2.6: The strength of support for our candidate seems to depend on whether we are interviewing men or women. Fifty percent of male voters support our candidate, but only 30% of women. What percent of women favor some other candidate? If we select a woman and a man at random and ask which candidate they support, in what percentage of cases do you think both will say they support our candidate?

Exercise 2.7: Would your answer to the previous question be the same if the man and the woman were coworkers? family members? fellow Republicans?

Exercise 2.8: Which approach do you think would be preferable to use in customer-satisfaction survey? To ask customers if they were or were not satisfied? Or to ask them to specify their degree of satisfaction on a 5-point scale? Why?

2.2.1 Permutations and Rearrangements

What is the probability of throwing exactly five heads in 10 tosses of a coin? The answer to this last question requires we understand something about permutations and combinations or rearrangements, a concept that will be extremely important in succeeding chapters.

Suppose we have three horses in a race. Call them A, B, and C. A could come in first, B could come in second, and C would be last. ABC is one possible outcome or permutation. But so are ACB, BAC, BCA, CAB, CBA; that is, there are six possibilities or *permutations* in all. Now suppose we have a nine-horse race. To determine the total number of possible orders of finish, we could enumerate them all one by one or we could use the following trick: We choose a winner (nine possibilities); we choose a second place finisher (eight remaining possibilities), and so forth until all positions are assigned. A total of $9! = 9 \times 8 \times 7 \times 6 \times 5 \times 4 \times 3 \times 2 \times 1$ possibilities in all. Had there been N horses in the race, there would have been $N!$ possibilities. $N!$ is read "N factorial."

Using R

```
> factorial(N)
```

yields the value of N! for any integer N.

Normally in a horse race, all our attention is focused on the first three finishers. How many possibilities are there? Using the same reasoning, it is easy to see there are $9 \times 8 \times 7$ possibilities or $9!/6!$. Had there been N horses in the race, there would have been $N!/(N-3)!$ possibilities.

Suppose we ask a slightly different question: In how many different ways can we select three horses from nine entries without regard to order, (i.e., we don't care which comes first, which second, or which third). This would be important if we had bet on a particular horse to show, that is, to be one of the first three finishers in a race.

In the previous example, we distinguished between first, second, and third place finishers; now, we're saying the order of finish doesn't make any difference. We already know there are $3! = 3 \times 2 \times 1 = 6$ different permutations of the three horses that finish in the first three places. So we take our answer to the preceding question $9!/6!$ and divide this answer in turn by $3!$. We write the result as

$$\binom{9}{3},$$

which is usually read as 9 choose 3. Note that

$$\binom{9}{3} = \binom{9}{6}.$$

In how many different ways can we assign nine cell cultures to two unequal experimental groups, one with three cultures and one with six? This would be the case if we had nine labels and three of the labels read "vitamin E" while six read "controls." If we could distinguish the individual labels, we could assign them in $9!$ different ways. But the order they are assigned within each of the experimental groups, $3!$ ways in the first instance and $6!$ in the other, won't affect the results. Thus, there are only $9!/6!3!$ or

$$\binom{9}{3} = 84 \text{ distinguishable ways.}$$

We can generalize this result to show the number of distinguishable ways N items can be assigned to two groups, one of k items, and one of $N - k$ is

$$\frac{N!}{k!(N-k)!} = \binom{N}{k}.$$

Using R, we would write

> choose(N,k).

What if we were to divide these same nine cultures among three equal-sized experimental groups? Then we would have $9!/3!3!3!$ distinguishable ways or *rearrangements*, written as

$$\binom{9}{3 \quad 3 \quad 3}.$$

Exercise 2.9: What is the value of 4!

Exercise 2.10: In how many different ways can we divide eight subjects into two equal-sized groups?

*2.2.2 Programming Your Own Functions in R

To tell whether our answer to question 2.10 is correct, we can program the factorial function in R.

```
➤ fact = function (num) {
+ prod(1:num)
+ }
```

Here, "1:num" yields a sequence of numbers from 1 to num, and **prod**() gives the product of these numbers.

To check your answer to the exercise, just type **fact**(4) after the >. Note that this function only makes sense for positive integers, 1, 2, 3, and so forth. To forestall errors, here is an improved version of the program, in which we print solution messages whenever an incorrect entry is made:

```
➤ fact = function (num) {
    if(num!=trunc(num))return(print("argument must be
      whole number"))
    if(num<0) return(print ("argument must be
      nonnegative integer"))
    else if(n= =0)return(1)
    else return(prod(1:num))
    }
```

Note that != means not equal to, while = = means equivalent to. Warning: writing if (n = 0) by mistake would lead to setting n equal to zero. Always use the = = in an if statement.

In general, to define a function in R, we use the form

```
function_name = function (param1, param2, . . .){
  . . .
  return (answer)
  }
```

If we want the function to do different things depending on satisfying a certain condition, we write

```
➤ if (condition) statement1 else statement2.
```

Possible conditions include $X > 3$ and name = = "Mike." Note that in R, the symbol = = means "is equivalent to," while the symbol = means "assign the value of the expression on the right of the equals sign to the variable on the left."

*Exercise 2.11: Write an R function that will compute the number of ways we can choose *m* from n things. Be sure to include solution messages to avoid problems.

2.2.3   Back to the Binomial

We used horses in this last example, but the same reasoning can be applied to coins or survivors in a clinical trial.* What is the probability of five heads in 10 tosses? What is the probability that five of 10 breast cancer patients will still be alive after 6 months?

We answer these questions in two stages. First, what is the number of different ways we can get five heads in 10 tosses? We could have thrown HHHHHTTTTT or HHHHTHTTTT, or some other combination of five heads and five tails for a total of 10 choose 5 or 10!/(5!5!) ways. The probability the first of these events occurs—five heads followed by five tails—is $(1/2)^{10}$. Combining these results yields

$$\Pr\{5 \text{ heads in 10 throws of a fair coin}\} = \binom{10}{5}(1/2)^{10}.$$

We can generalize the preceding to an arbitrary probability of success p, $0 \le p \le 1$. The probability of failure is $1 - p$. The probability of k successes in n trials is given by the binomial formula

$$\binom{n}{k}(p)^k(1-p)^{n-k} \quad \text{for } 0 \le k \le n. \tag{2.1}$$

Exercise 2.12: What is the probability of getting at least one head in six flips of a fair coin?

2.2.4   The Problem Jury

At issue in Ballew v. Georgia[†] brought before the Supreme Court in 1978 was whether the all-white jury in Ballew's trial represented a denial of Ballew's rights.[‡] In the 1960s and 1970s, U.S. courts held uniformly that the use of race, gender, religion, or political affiliation to bar citizens from jury service would not be tolerated. In one case in 1963 in which I assisted the defense on appeal, we were able to show that only one black had served on some 163 consecutive jury panels. In this case, we were objecting—successfully—to the methods used to select the jury. In *Ballew*, the defendant was not objecting to the methods but to the composition of the specific jury that had judged him at trial.

In the district in which Ballew was tried, blacks comprised 10% of the population, but Ballew's jury was entirely white. Justice Blackmun wanted to know what

---

* If, that is, the probability of survival is the same for every patient. When there are obvious differences from trial to trial—for example, one subject is an otherwise healthy 35-year old male, the other an elderly 89-year old who has just recovered from pneumonia—this simple binomial model would not apply.

[†] 435 U.S. 223, 236–237 (1978).

[‡] Strictly speaking, it is not the litigant but the potential juror whose rights might have been interfered with. For more on this issue, see Chapter 2 of Phillip Good, *Applications of Statistics in the Courtroom* (Chapman and Hall, 2001).

the probability was that a jury of 12 persons selected from such a population in accordance with the law would fail to include members of the minority.

If the population in question is large enough, say a hundred thousand or so, we can assume that the probability of selecting a nonminority juryperson is a constant 90 out of 100. The probability of selecting two nonminority persons in a row according to the product rule for independent events is 0.9×0.9 or 0.81. Repeating this calculation 10 more times, once for each of the remaining 10 jurypersons, we get a probability of $0.9 \times 0.9 \times \ldots \ldots \times 0.9 = 0.28243$ or 28%.

Not incidentally, Justice Blackmun made exactly this same calculation and concluded that Ballew had not been denied his rights.

*2.3 CONDITIONAL PROBABILITY

Conditional probability is one of the most difficult of statistical concepts, not so much to understand as to accept in all its implications. Recall that mathematicians arbitrarily assign a probability of 1 to the result that something will happen—the coin will come up heads or tails, and 0 to the probability that nothing will occur. But real life is more restricted: a series of past events has preceded our present, and every future outcome is conditioned on this past. Consequently, we need a method whereby the probabilities of just the remaining possibilities sum to 1.

We define the *conditional probability* of an event A given another event B, written $P(A|B)$, to be the ratio $P(A \text{ and } B)/P(B)$. To show how this would work, suppose we are playing Craps, a game in which we throw two six-sided die. Clearly, there are $6 \times 6 = 36$ possible outcomes. One (and only one) of these 36 outcomes is snake eyes, a 1 and a 1.

Now, suppose we throw one die at a time (a method that is absolutely forbidden in any real game of Craps, whether in the Bellagio or a back alley) and a 1 appears on the first die. The probability that we will now roll snake eyes, that is, that the second die will reveal a 1 also, is 1 out of 6 possibilities or $(1/36)/(1/6) = 6/36 = 1/6$.

The conditional probability of rolling a total of seven spots on the two dice is 1/6. And the conditional probability of the spots on the two die summing to 11, another winning combination, is 0. Yet before we rolled the two dice, the unconditional probability of rolling snake eyes was 1 out of 36 possibilities, and the probability of 11 spots on the two die was 2/36th (a 5 and a 6 or a 6 and a 5).

Now, suppose I walk into the next room where I have two decks of cards. One is an ordinary deck of 52 cards, half red and half black. The other is a trick deck in which all the spots on the cards are black. I throw a coin—I'm still in the next room so you don't get to see the result of the coin toss—and if the coin comes up heads I stick the trick deck in my pocket, otherwise I take the normal deck. Now, I come back into the room and offer to show you a card chosen at random from the deck in my pocket. The card has black spots. Would you like to bet on whether or not I'm carrying the trick deck?

[STOP: Think about your answer before reading further.]

Common sense would seem to suggest that the odds are still only 50–50 that it's the trick deck I'm carrying. You didn't really learn anything from seeing a card that could have come from either deck. Or did you?

Let's use our conditional probability relation to find out whether the odds have changed. First, what do we know? As the deck was chosen at random, we know that the probability of the card being drawn from the trick deck is the same as the probability of it being drawn from the normal one:

$$P(T^c) = P(T) = \frac{1}{2}.$$

Here, T denotes the event that I was carrying a trick deck and T^c denotes the complementary event (not T) that I was carrying the normal deck.

We also know two conditional probabilities. The probability of drawing a black card from the trick deck is, of course, 1, while that of drawing a black card from a deck which has equal numbers of black and red cards is 1/2. In symbols, $P(B|T) = 1$ and $P(B|T^c)$ is 1/2.

What we'd like to know is whether the two conditional probabilities $P(T|B)$ and $P(T^c|B)$ are equal. We begin by putting the two sets of facts we have together, using our conditional probability relation, $P(B|T) = P(T \text{ and } B)/P(T)$.

We know two of the values in the first relation, $P(B|T)$ and $P(T)$, and so we can solve for $P(B \text{ and } T) = P(B|T)P(T) = 1 \times 1/2$. Similarly,

$$P(B \text{ and } T^c) = P(B|T^c)P(T^c) = \frac{1}{2} \times \frac{1}{2} = \frac{1}{4}.$$

Take another look at our Venn diagram in Figure 2.1. All the events in outcome B are either in A or in its complement A^c. Similarly,

$$P(B) = P(B \text{ and } T) + P(B \text{ and } T^c) = \frac{1}{2} + \frac{1}{4} = \frac{3}{4}.$$

We now know all we need to know to calculate the conditional probability P(T|B) for our conditional probability relation can be rewritten to interchange the roles of the two outcomes, giving

$$P(T|B) = P(B \text{ and } T)/P(B) = \frac{1}{2} \bigg/ \frac{3}{4} = \frac{2}{3}.$$

By definition $P(T^c|B) = 1 - P(T|B) = 1/3 < P(T|B)$.

The odds have changed. Before I showed you the card, the probability of my showing you a black card was

$$1 \times \frac{1}{2} + \frac{1}{2} \times \frac{1}{2} \text{ or } \frac{3}{4}.$$

When I showed you a black card, the probability it came from a black deck was ½ divided by ¾ or 2/3!

Exercise 2.13: If R denotes a red card, what would be $P(T|R)$ and $P(T^c|R)$?

> ### A Too Real Example
>
> Think the previous example was artificial? That it would never happen in real life? My wife and I just came back from a car trip. On our way up the coast, I discovered that my commuter cup leaked, but, desperate for coffee, I wrapped a towel around the cup and persevered. Not in time, my wife noted, pointing to the stains on my jacket.
>
> On our way back down, I lucked out and drew the cup that didn't leak. My wife congratulated me on my good fortune and then, ignoring all she might have learned had she read this text, proceeded to drink from the remaining cup! So much for her new Monterey Bay Aquarium sweatshirt.

2.3.1 Market Basket Analysis

Many supermarkets collect data on purchases using barcode scanners located at the check-out counter. Each transaction record lists all items bought by a customer on a single purchase transaction. Executives want to know whether certain groups of items are consistently purchased together. They use these data for adjusting store layouts (placing items optimally with respect to each other), for cross-selling, for promotions, for catalog design and to identify customer segments based on buying patterns.

If a supermarket database has 100,000 point-of-sale transactions, out of which 2000 include both items A and B and 800 of these include item C, the *association rule* "If A and B are purchased then C is purchased on the same trip" has a *support* of 800 transactions (alternatively $0.8\% = 800/100,000$) and a *confidence* of 40% ($= 800/2000$).

> **Exercise 2.14:** Suppose you have the results of a market-basket analysis in hand. (1) If you wanted an estimate of the probability that a customer will purchase anchovies, would you use the support or the confidence? (2) If you wanted an estimate of the probability that a customer with anchovies in her basket will also purchase hot dogs, would you use the support or the confidence?

2.3.2 Negative Results

Suppose you were to bet on a six-horse race in which the horses carried varying weights on their saddles. As a result of these handicaps, the probability that a specific horse will win is exactly the same as that of any other horse in the race. What is the probability that your horse will come in first?

Now suppose, to your horror, a horse other than the one you bet on finishes first. No problem; you say, "I bet on my horse to place," that is, you bet it would come in first **or** second. What is the probability you still can collect on your ticket? That is, what is the conditional probability of your horse coming in second when it did not come in first?

One of the other horses did finish first, which leaves five horses still in the running for second place. Each horse, including the one you bet on, has the same probability to finish second, so the probability you can still collect is one out of five. Agreed?

Just then, you hear the announcer call out that the horses are about to line up for the second race. Again there are six horses and each is equally likely to finish first. What is the probability that if you bet on a horse to place in the second race that you will collect on your bet? Is it 1/6 + 1/5?

There are three ways we can arrive at the correct answer when all horses are equal in all respects:

1. We could notice that the probability that our horse will finish second is exactly the same as the probability that it will finish first (or the probability that it will finish dead last, for that matter). As these are mutually exclusive outcomes, their probabilities may be added. The probability is 2/6th that your horse finishes first or second.

2. We could list all 6! mutually exclusive outcomes of the race and see how many would lead to our collecting on our bet—but this would be a lot of work.

3. Or, we could trace the paths that lead to the desired result. For example, either our horse comes in first with probability 1/6 or it does not, with probability 5/6. If it doesn't come in first, it might still come in second with probability 1/5. The overall probability of your collecting on your bet is Pr{your horse wins} + Pr{your horse doesn't win}*Pr{your horse is first among the nonwinning horses} = 1/6 + 5/6*1/5 = 2/6.

Exercise 2.15: Suppose 10 people are in a class. What is the probability that no two of them were born on the same day of the week? What is the probability that all of them were born in nonoverlapping four-week periods?

Exercise 2.16: A spacecraft depends on five different mission-critical systems. If any of these systems fail, the flight will end in catastrophe. Taken on an individual basis, the probability that a mission-critical system will fail during the flight is 1/10. (1) What is the probability that the flight will be successful?

NASA decides to build in redundancies. Every mission-critical system has exactly one back-up system that will take over in the event that the primary system fails. The back-up systems have the same probability of failure as the primaries. (2) What is the probability that the flight will be successful?

Exercise 2.17: A woman sued a Vegas casino alleging the following: she asked a security guard to hold her slot machine while she hit the buffet; he let somebody else use "her" machine while she was gone; that "somebody else" hit the jackpot; that jackpot was rightfully hers. The casino countered that jackpots were triggered by a random clock keyed to the 1/1000th of a second, thus, even had the woman been playing the machine continuously, she might not have hit the jackpot. How would you rule if you were a judge?

Exercise 2.18: In the United States in 1985, there were 2.1 million deaths from all causes, compared with 1.7 million in 1960. Does this mean it was safer to live in the United States in the 1960s than in the 1980s?

Exercise 2.19: You are a contestant on "Let's make a deal." Monty offers you a choice of three different curtains and tells you there is a brand new automobile behind one of them plus enough money to pay the taxes in case you win the car. You tell Monty you want to look behind curtain number 1. Instead, he throws back curtain number 2 to reveal . . . a child's toy. "Would you like to choose curtain number 3 instead of number 1?" Monty asks. Well, would you?

2.4 INDEPENDENCE

A key element in virtually all the statistical procedures we will consider in this text is that the selection of one member of a sample takes place *independently* of the selection of another. In discussing the game of Craps, we assumed that the spots displayed on the first die were *independent* of the spots displayed on the second. When statistics are used, either we:

1. Assume observations are independent.
2. Test for independence
3. Try to characterize the nature of the dependence (Chapter 7).

Two events or observations are said to be independent of one another, providing knowledge of the outcome or value of the one gives you no information regarding the outcome or value of the other.

In terms of conditional probabilities, two events A and B are independent of one another providing that $P(A|B) = P(A)$, that is, our knowledge that B occurred does not alter the likelihood of A. (This relation should be read as "the probability that A occurs given that B occurs is equal to the probability that A occurs.")

We can use this relation to show that if A and B are independent, then the probability they will *both* occur is the product of their separate probabilities, $P(A$ and $B) = P(A)*P(B)$. For from the definition of conditional probability, $P(A$ and $B) = P(A)*P(A$ and $B|A) = P(A)*P(B|A) = P(A)*P(B)$.

Warning: Whether events are independent of one another will depend upon the context. Imagine that three psychiatrists interview the same individual who we shall suppose is a paranoid schizophrenic. The interviews take place at different times and the psychiatrists are not given the opportunity to confer with each other either before or after the interviews take place.

Suppose now that these psychiatrists are asked for their opinions on (1) the individual's sanity, and, having been informed of the patient's true condition, (2) their views on paranoid schizophrenia. In the first case, their opinions will be independent of one another; in the second case, they will not.

Exercise 2.20: Can two independent events be mutually exclusive?

Exercise 2.21: Draw a Venn diagram depicting two independent events one of which is twice as likely to occur as the other.

Exercise 2.22: Do the following constitute independent observations?

  a. Several students sitting together at a table are asked who their favorite movie actress is.
  b. The number of abnormalities in each of several tissue sections taken from the same individual.
  c. Opinions of several individuals whose names you obtained by sticking a pin through a phone book, and calling the "pinned" name on each page.
  d. Opinions of an ardent Democrat and an ardent Republican.
  e. Today's price in Australian dollars of the German mark and the Japanese yen.

Exercise 2.23: Based on the results in the following *contingency tables*, would you say that sex and survival are independent of one another in Table A? In Table B?

Table A		
	Alive	Dead
Men	15	5
Women	15	10

Table B		
	Alive	Dead
Men	15	10
Women	12	8

Exercise 2.24: Provide an example in which an observation X is independent of the value taken by an observation Y, X is independent of a third observation Z, and Y is independent of Z, but X, Y, and Z are dependent.

2.5 APPLICATIONS TO GENETICS

All the information needed to construct an organism, whether a pea plant, a jelly fish, or a person, is encoded in its genes. Each gene contains the information needed to construct a single protein. Each of our cells has two copies of each gene, one obtained from our mother and one from our father. We will contribute just one of these copies to each of our offspring. Whether it is the copy we got from our father or the one from our mother is determined entirely by chance.

You could think of this as flipping a coin, one side says "mother's gene," the other side says "father's gene." Each time a sperm is created in our testis or an ovum in our ovary, the coin is flipped.

There may be many forms of a single gene; each such form is called an allele. Some alleles are defective, incapable of constructing the necessary protein. For example, my mother was rh–, meaning that both her copies of the rh gene were incapable of manufacturing the rh protein that is found in red blood cells. This also means that the copy of the rh gene I obtained from my mother was rh–. But my blood tests positive for the rh protein, which means that the rh gene I got from my father was rh+.

Exercise 2.25: The mother of my children was also rh–. What proportion of our children would you expect to be rh-?

Exercise 2.26: Sixteen percent of the population of the United States is rh–. What percentage do you expect to have at least one rh– gene? (Remember, as long as a person has even one rh+ gene, they can manufacture the rh protein.)

The gene responsible for making the A and B blood proteins has three alleles, A, B, and O. A person with two type 0 alleles will have blood type O. A person with one A allele and one B allele will have blood type AB. Only 4% of the population of the United States has this latter blood type.

Our genes are located on chromosomes. The chromosomes come in pairs, one member of each pair being inherited from the father and one from the mother. Your chromosomes are passed onto the offspring independently of one another.

Exercise 2. 27: The ABO and rh genes are located on different chromosomes and thus are inherited independently of one another. What percentage of the population of the United States would you expect to have the AB rh+ blood type?

2.6 SUMMARY AND REVIEW

In this chapter, we introduced the basics of probability theory and independence and applied the elements of both to genetics. We learned additional R commands for controlling program flow (if, else) and for displaying properties of binomial distributions. We also learned how to create our own special purpose R functions.

Exercise 2.28: Make a list of all the italicized terms in this chapter. Provide a definition for each one along with an example.

Exercise 2.29: (Read and reread carefully before even attempting an answer) A magician has three decks of cards, one with only red cards, one that is a normal deck, and one with only black cards. He walks into an adjoining room and returns with only a single deck. He removes the top card from the deck and shows it to you. The card is black. What is the probability that the deck from which the card came consists only of black cards?

Exercise 2.30: An integer number is chosen at random. What is the probability that it is divisible by 2? What is the probability that it is divisible by 17? What is the probability that it is divisible by 2 and 17? What is the probability that it is divisible by 2 or 17? (Hint: A Venn diagram would be a big aid in solving this last part.)

Exercise 2.31: Pete, Phil, and Myron are locked in a squash court after hours with only a Twinkie and a coin between them. The only thing all three can agree on is that they want a whole Twinkie or nothing. Myron suggests that Pete and Phil flip the coin, and that the winner flips a coin with him to see who gets the Twinkie. Phil who is a graduate student in statistics says this is unfair. Is it unfair and why? How would you decide who gets the Twinkie?

Exercise 2.32: People v. Collins,* concerned a prosecution of an African-American and his Caucasian wife for robbery. The victim testified her purse was snatched by a girl with a blond ponytail; a second witness testified he saw a blond girl, ponytail flying, enter a yellow convertible being driven by an African-American with a beard and mustache. Neither witness could identify the suspects directly. In an attempt to prove the defendants were in fact the persons who had committed the crime, the prosecutor called a college instructor of mathematics to establish that, assuming the robbery were committed by a Caucasian woman with a blond ponytail who left the scene in a yellow Lincoln accompanied by a African-American with a beard and mustache, there was an overwhelming probability the crime was committed by any couple answering to such distinctive characteristics.

Assume you were on the jury. What would you decide and why?

---

* 68 Cal. 2d 319, 66 Cal. Rptr. 497 (1968).

Chapter 3

Two Naturally Occurring Probability Distributions

In this chapter, you learn to recognize and describe the binomial and Poisson probability distributions in terms of their *cumulative distribution functions* and their *expectations*. You'll learn when to apply these distributions in your future work along with methods for estimating their parameters using R.

3.1 DISTRIBUTION OF VALUES

Life constantly calls upon us to make decisions. Should penicillin or erythromycin be prescribed for an infection? Is one brand of erythromycin preferable to another? Which fertilizer should be used to get larger tomatoes? Which style of dress should our company manufacture (i.e., which style will lead to greater sales)?

My wife often asks me what appears to be a very similar question, "which dress do you think I should wear to the party?" But this question is really quite different from the others as it asks what a specific individual, me, thinks about dresses that are to be worn by another specific individual, my wife. All the other questions reference the behavior of a yet-to-be-determined individual selected *at random* from a population.

Is Alice taller than Peter? It won't take us long to find out: We just need to put the two back to back or measure them separately.* The somewhat different question, "are girls taller than boys?" is not answered quite so readily. How old are the boys and girls? Generally, but not always, girls are taller than boys until they enter adolescence. Even then, what may be true in general for boys and girls of a specified age group may not be true for a particular girl and a particular boy.

Put in its most general and abstract form, what we are asking is whether a numerical observation X made on an individual drawn at random from one population will be larger than a similar numerical observation Y made on an individual drawn at random from a second population.

* Quick: Which methods are likely to yield the more accurate comparison? Why?

Introduction to Statistics Through Resampling Methods and R, Second Edition. Phillip I. Good.
© 2013 John Wiley & Sons, Inc. Published 2013 by John Wiley & Sons, Inc.

3.1.1 Cumulative Distribution Function

Let $F_W[w]$ denote the probability that the numerical value of an observation from the distribution W will be less than or equal to w. $F_W[w]$ is a monotone nondecreasing function, that is, a graph of F_W either rises or remains level as w increases. In symbols, if $\varepsilon > 0$, then $0 = F_W[-\infty] \leq F_W[w] = \Pr\{W \leq w\} \leq \Pr\{W \leq w + \varepsilon\} = F_W[w + \varepsilon] \leq F_W[\infty] = 1$.

Two such *cumulative distribution functions*, F_w and G_w, are depicted in Figure 3.1.

Note the following in this figure:

1. Both distributions start at 0 and end at 1
2. F_X is to the left of G_X. As can be seen by drawing lines perpendicular to the value axis, $F_X[x] > G_X[x]$ for all values of X. As can be seen by drawing lines perpendicular to the percentile-axis, all the percentiles of the cumulative distribution G are greater in value than the percentiles of the cumulative distribution F.
3. Most of the time an observation taken at random from the distribution F_X will be smaller than if it is taken from the distribution G_X.
4. Still, there is a nonzero probability that an observation from F_X will be larger than one from G_X.

Many treatments act by shifting the distribution of values, as shown in Figure 3.1. The balance of this chapter and the next is concerned with the detection of such treatment effects. The possibility exists that the effects of treatment are more complex than is depicted in Figure 3.1. In such cases, the introductory methods described in

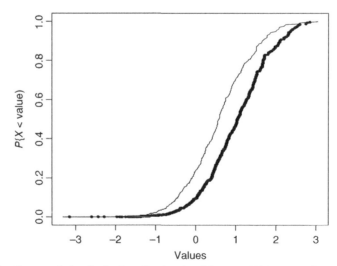

Figure 3.1 Two cumulative distributions F and G that differ by a shift in the median value. The fine line denotes F; the thick line corresponds to G.

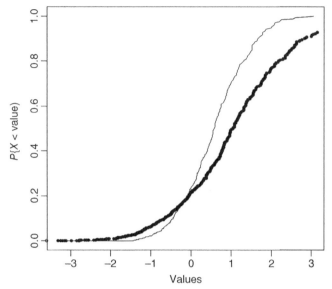

Figure 3.2 Two cumulative distributions whose mean and variance differ.

this text may not be immediately applicable; see, for example, Figure 3.2, in which the population represented by the thick line is far more variable than the population represented by the thin line.

> **Exercise 3.1:** Is it possible that an observation drawn at random from the distribution F_X depicted in Figure 3.1 could have a value larger than an observation drawn at random from the distribution G_X?*

3.1.2 Empirical Distribution Function

Suppose we've collected a sample of n observations $X = \{x_1, x_2, \ldots, x_n\}$. The *empirical* (observed) *cumulative distribution function* $F_n[x]$ is equal to the number of observations that are less than or equal to x divided by the size of our sample n. If we've sorted the sample using the R command sort(X) so that $x[1] \le x[2] \le \ldots \le x[n]$, then the empirical distribution function $F_n[x]$ is

```
0     if x < x1
k/n   if x lies between x[k] and x[k+1]
1     if x > x[n].
```

If the observations in the sample all come from the same population distribution **F** and are independent of one another, then as the sample size n gets larger, F_n will begin to resemble **F** more and more closely. We illustrate this point in Figure 3.3, with samples of size 10, 100, and 1000 all taken from the same distribution.

---

* If the answer to this exercise is not immediately obvious—you'll find the correct answer at the end of this chapter—you should reread Chapters 1 and 2 before proceeding further.

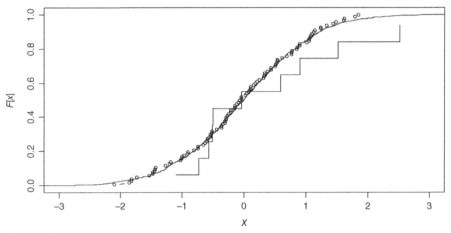

Figure 3.3   Three empirical distributions based on samples of size 10 (staircase), 100 (circles), and 1000 independent observations from the same population. The R code needed to generate this figure will be found in Chapter 4.

Figure 3.3 reveals what you will find in your own samples in practice: The fit between the empirical (observed) and the theoretical distribution is best in the middle of the distribution near the median and worst in the tails.

3.2 DISCRETE DISTRIBUTIONS

We need to distinguish between *discrete* random observations made when recording the number of events that occur in a given period, from the *continuous* random observations made when taking measurements.

Discrete random observations usually take only integer values (positive or negative) with nonzero probability. That is,

$$\Pr\{X = k\} = f_k \quad \text{if } k = \{\ldots -1, 0, 1, 2, \ldots\}$$
$$= 0 \quad \text{otherwise.}$$

The cumulative distribution function $F[x]$ is

$$\sum_{k \le x} f_k.$$

The expected value of such a distribution is defined as

$$\sum_{k \le x} k f_k.$$

The binomial and the Poisson are discrete distributions that are routinely encountered in practice.

Exercise 3.2: If X and Y are two independent discrete random variables, show that the expected value of $X + Y$ is equal to the sum of the expected values of X and Y. Hint: Use the knowledge that

$$\Pr\{X + Y = k\} = \sum_{j=0}^{k} \Pr\{X = j \text{ and } Y = k - j\}.$$

We can also show that the variance of two independent discrete random variables is the sum of their variances.

3.3 THE BINOMIAL DISTRIBUTION

The binomial distribution arises whenever we perform a series of independent trials each of which can result in either success or failure, and the probability of success is the same for each trial. Survey questions that offer just two alternatives, male or female, smoker or nonsmoker, like or dislike, give rise to binomial distributions.

Recall from the preceding chapter (Section 2.2.3) that when we perform n trials (flip a coin n times, mark n completed survey questions), each with a probability p of success, the probability of observing n successes is given by the following formula:

$$\Pr\{X = k | n, p\} = \binom{n}{k}(p)^k (1 - p)^{n-k} \quad \text{for } k = \{0, 1, 2, \ldots, n\}$$
$$= 0 \qquad \qquad \text{otherwise.}$$

where $\binom{n}{k}$ denotes the number of different possible rearrangements of k successes and $n - k$ failures. The cumulative distribution function of a binomial is a step function, equal to zero for all values less than 0 and to one for all values greater than or equal to n.

The expected number of successes in a large number of repetitions of a single binomial trial is p. The expected number of successes in a sufficiently large number of sets of n binomial trials each with a probability p of success is np.

*3.3.1 Expected Number of Successes in n Binomial Trials

Though it seems obvious that the mean of a sufficiently large number of sets of n binomial trials each with a probability p of success will be equal to np, many things that seem obvious in mathematics aren't. Besides, if it's that obvious, we should be able to prove it. Given sufficient familiarity with high school algebra, it shouldn't be at all difficult.

If a variable X takes a discrete set of values $\{\ldots, 0, 1, \ldots, k, \ldots\}$ with corresponding probabilities $\{\ldots f_0, f_1, \ldots, f_k, \ldots\}$, its mean or *expected value*, written EX, is given by the summation: $\ldots + 0f_0 + 1f_1 + \ldots + kf_k + \ldots$ which we may also write as $\Sigma_k kf_k$. For a binomial variable, this sum is

$$\text{EX} = \sum_{k=0}^{n} k \binom{n}{k} (p)^k (1-p)^{n-k}$$

$$= \sum_{k=0}^{n} k \frac{n(n-1)!}{k!(n-k)!} p(p)^{k-1} (1-p)^{n-1-(k-1)}.$$

Note that the term on the right is equal to zero when $k = 0$, so we can start the summation at $k = 1$. Factoring n and p outside the summation and using the k in the numerator to reduce $k!$ to $(k-1)!$, we have

$$\text{EX} = np \sum_{k=1}^{n} \frac{(n-1)!}{(k-1)![(n-1)-(k-1)]!} (p)^{k-1} (1-p)^{(n-1)-(k-1)}.$$

If we change the notation a bit, letting $j = (k-1)$ and $m = (n-1)$, this can be expressed as

$$\text{EX} = np \sum_{j=0}^{m} \binom{m}{j} (p)^j (1-p)^{m-j}.$$

The summation on the right side of the equals sign is of the probabilities associated with *all* possible outcomes for the binomial random variable $B(n, p)$, so it must be and is equal to 1. Thus, $\text{EX} = np$, a result that agrees with our intuitive feeling that in the long run, the number of successes should be proportional to the probability of success.

3.3.2   Properties of the Binomial

Suppose we send out several hundred individuals to interview our customers and find out if they are satisfied with our products. Each individual has the responsibility of interviewing exactly 10 customers. Collating the results, we observe several things:

- 740 out of every 1000 customers reported they were satisfied.
- Results varied from interviewer to interviewer.
- About 6% of the samples included no dissatisfied customers
- A little more than 2% of the samples included 6 or more dissatisfied customers
- The median number of satisfied customers per sample was 7
- The modal number of satisfied customers per sample was 8.

When we reported these results to our boss, she only seemed interested in the first of them. "Results always vary from interviewer to interviewer, and from sample to sample. And the percentages you reported, apart from the 74% satisfaction rate, are immediate consequences of the binomial distribution."

Clearly, our boss was familiar with the formula for k successes in n trials given in Section 2.2.3. From our initial finding, she knew that $p = 0.74$. Thus,

$$\Pr\{k \text{ satisfied customers in sample of } 10\} = \binom{10}{k}(0.74)^k(0.26)^{n-k} \text{ for } 0 \le k \le 10.$$

To find the median of this distribution, its 50th percentile, we can use R as follows:

```
➤ qbinom(.50,10,0.74)
[1]  7
```

To find the proportion of samples with no dissatisfied customers:

```
➤ dbinom(10,10,0.74)
[1]  0.0492399
```

To find the proportion of samples with four or less satisfied customers:

```
➤ pbinom(4,10,0.74)
```

To display this binomial distribution completely and find its mode:

```
➤ #create a vector called binom.vals to hold all the
  #possible values the results of 10 binomial trials
  #might take.
  #store 0 to 10 in the vector to binom.vals
➤ binom.vals = 0:10
➤ binom.prop = dbinom(binom.vals,10,0.74)
➤ #display the results
➤ binom.prop
 [1] 1.411671e-06 4.017833e-05 5.145917e-04 3.905619e-
  03 1.945299e-02
 [6] 6.643943e-02 1.575807e-01 2.562851e-01 2.735350e-
  01 1.730051e-01
[11] 4.923990e-02
```

The scientific notation used by R in reporting very small or very large values may be difficult to read. Try

```
➤ round(binom.prop,3)
```

for a slightly different view. The 3 asks for three places to the right of the decimal point. Any road, 4.924e-02 is the same as 0.04924.

To find the mode, proceed in two steps

```
➤ max(binom.prop)
[1]  2.735350e-01
```

The elements of vectors are numbered consecutively beginning at [1]. The mode of the distribution, the outcome with the greatest proportion show above, is located at position 9 in the vector.

```
➢ binom.vals[9]
[1]  8
```

tells us that the modal outcome is 8 successes.

To find the mean or *expected value* of this binomial distribution, let us first note that the computation of the arithmetic mean can be simplified when there are a large number of ties by multiplying each distinct number k in a sample by the frequency f_k with which it occurs;

$$\bar{X} = \sum_k k f_k.$$

We can have only 11 possible outcomes as a result of our interviews: 0, 1, . . . , or 10 satisfied customers. We know from the binomial distribution the frequency f_i with which each outcome may be expected to occur; so that the population mean is given by the formula

$$\sum_{i=0}^{10} i \binom{10}{i} (p)^i (1-p)^{10-i}.$$

For our binomial distribution, we have the values of the variable in the vector binom. vals and their proportions in binom.prop. To find the expected value of our distribution using R, type

```
➢ sum(binom.vals*binom.prop)
[1]  7.4
```

This result, we notice, is equal to 10*0.74 or 10p. This is not surprising, as we know that the expected value of the binomial distribution is equal to the product of the number of trials and the probability of success at each trial.

Warning: In the preceding example, we assumed that our sample of 1000 customers was large enough that we could use the proportion of successes in that sample, 740 out of 1000 as if it were the true proportion in the entire distribution of customers. Because of the variation inherent in our observations, the true proportion might have been greater or less than our estimate.

Exercise 3.3: Which is more likely: Observing two or more successes in 8 trials with a probability of one-half of observing a success in each trial, or observing three or more successes in 7 trials with a probability of 0.6 of observing a success? Which set of trials has the greater expected number of successes?

Exercise 3.4: Show without using algebra that if X and Y are independent identically distributed binomial variables $B(n,p)$, then $X + Y$ is distributed as $B(2n,p)$.

Unless we have a large number of samples, the observed or *empirical distribution* may differ radically from the expected or theoretical distribution.

Histogram of binom

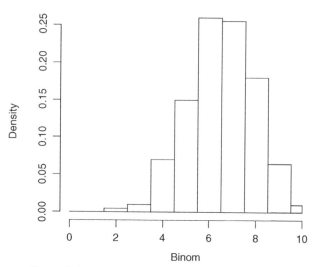

Figure 3.4 Histogram, based on 200 trials of $B(10,0.65)$.

Exercise 3.5: One can use the R function rbinom to generate the results of binomial sampling. For example, binom = rbinom(100,10,0.6) will store the results of 100 binomial samples of 10 trials, each with probability of 0.6 of success in the vector binom. Use R to generate such a vector of sample results and to produce graphs to be used in comparing the resulting empirical distribution with the corresponding theoretical distribution.

Exercise 3.6: If you use the R code hist(rbinom(200,10,0.65)), the resulting histogram will differ from that depicted in Figure 3.4. Explain why.

3.4 MEASURING POPULATION DISPERSION AND SAMPLE PRECISION

The *variance* of a sample, a measure of the variation within the sample, is defined as the sum of the squares of the deviations of the individual observations about their mean divided by the sample size minus 1. In symbols, if our observations are labeled X_1, X_2, up to X_n, and the mean of these observations is written as $\bar{X}$, then the variance of our observations is equal to

$$\frac{1}{(n-1)} \sum_{i=1}^{n} (X_i - \bar{X})^2.$$

Both the mean and the variance will vary from sample to sample. If the samples are very large, there will be less variation from sample to sample, and the sample mean and variance will be very close to the mean and variance of the population.

Sidebar: Why n – 1?

Why do we divide the sum of the deviations about the mean by $n - 1$, rather than n? When our sample consists of a single observation, $n = 1$, there is no variation. When we have two observations,

$$\sum_{i=1}^{2} (X_i - \bar{X})^2 = (X_1 - \bar{X})^2 + (X_2 - \bar{X})^2.$$

But

$$(X_1 - \bar{X})^2 = \left(X_1 - \frac{1}{2}(X_1 + X_2) \right)^2 = \frac{1}{4}(X_1 - X_2)^2 = \frac{1}{4}(X_2 - X_1)^2 = (X_2 - \bar{X})^2.$$

That is, we really only have one distinct squared deviation.

Exercise 3.7: What is the sum of the deviations of the observations from their arithmetic mean? That is, what is

$$\sum_{i=1}^{n} (X_i - \bar{X})?$$

*Exercise 3.8: If you know calculus, use the result of Exercise 3.7 to show that

$$\frac{1}{(n-1)} \sum_{i=1}^{n} (X_i - \mu)^2,$$

is minimized if we chose $\mu = \bar{X}$. In other words, the sum of the squared deviations is a minimum about the mean of a sample.

The problem with using the variance as a measure of dispersion is that if our observations, on temperature, for example, are in degrees Celsius, then the variance would be expressed in square degrees, whatever these are. More often, we report the *standard deviation* σ, the square root of the variance, as it is in the same units as our observations.

The standard deviation of a sample also provides a measure of its precision.

Exercise 3.9: Recall the classroom data of Chapter 1, classdata = c(141, 156.5, 162, 159, 157, 143.5, 154, 158, 140, 142, 150, 148.5, 138.5, 161, 153, 145, 147, 158.5, 160.5, 167.5, 155, 137). Compute the data's variance and standard deviation, using the R functions **var**() and **sqrt**().

The variance of a binomial variable B(n,p) is $np(1 - p)$. Its standard deviation is $\sqrt{np(1-p)}$. This formula yields an important property that we will make use of in a latter chapter when we determine how large a sample to take. For if the precision of a sample is proportional to the square root of its sample size, then to make estimates that are twice as precise, we need to take four times as many observations.

3.5 POISSON: EVENTS RARE IN TIME AND SPACE

The decay of a radioactive element, an appointment to the U.S. Supreme Court, and a cavalry officer kicked by his horse have in common that they are relatively rare but inevitable events. They are inevitable, that is, if there are enough atoms, enough seconds or years in the observation period, and enough horses and momentarily careless men. Their frequency of occurrence has a Poisson distribution.

The number of events in a given interval has the Poisson distribution if

a. it is the cumulative result of a large number of independent opportunities, each of which has only a small chance of occurring; and
b. events in nonoverlapping intervals are independent.

The intervals can be in space or time. For example, if we seed a small number of cells into a Petri dish that is divided into a large number of squares, the distribution of cells per square follows the Poisson. The same appears to be true in the way a large number of masses in the form of galaxies are distributed across a very large universe.

Like the binomial variable, a Poisson variable only takes non-negative integer values. If the number of events X has a Poisson distribution such that we may expect an average of λ events per unit interval, then $\Pr\{X = k\} = \lambda^k e^{-\lambda}/k!$ for $k = 0, 1, 2, \ldots$. For the purpose of testing hypotheses concerning λ as discussed in the chapter following, we needn't keep track of the times or locations at which the various events occur; the number of events k is *sufficient*.

Exercise 3.10: If we may expect an average of two events per unit interval, what is the probability of no events? Of one event?

Exercise 3.11: Show without using algebra that the sum of a Poisson with the expected value λ_1 and a second independent Poisson with the expected value λ_2 is also a Poisson with the expected value $\lambda_1 + \lambda_2$.

3.5.1 Applying the Poisson

John Ross of the Wistar Institute held there were two approaches to biology: the analog and the digital. The analog was served by the scintillation counter: one ground up millions of cells then measured whatever radioactivity was left behind in the stew after centrifugation; the digital was to be found in cloning experiments where any necessary measurements would be done on a cell-by-cell basis.

John was a cloner, and, later, as his student, so was I. We'd start out with 10 million or more cells in a 10-mL flask and try to dilute them down to one cell per mL. We were usually successful in cutting down the numbers to 10,000 or so. Then came the hard part. We'd dilute the cells down a second time by a factor of 1:10 and hope we'd end up with 100 cells in the flask. Sometimes we did. Ninety percent of the time, we'd end up with between 90 and 110 cells, just as the Poisson distribution predicted. But just because you cut a mixture in half (or a dozen, or a 100 parts) doesn't mean you're going to get equal numbers in each part. It means the probability of getting a particular cell is the same for all the parts. With large numbers of cells, things seem to even out. With small numbers, chance seems to predominate.

Things got worse, when I went to seed the cells into culture dishes. These dishes, made of plastic, had a rectangular grid cut into their bottoms, so they were divided into approximately 100 equal size squares. Dropping 100 cells into the dish meant an average of 1 cell per square. Unfortunately for cloning purposes, this average didn't mean much. Sometimes, 40% or more of the squares would contain two or more cells. It didn't take long to figure out why. Planted at random, the cells obey the Poisson distribution in space. An average of one cell per square means

$\Pr\{$No cells in a square$\} = 1 * e^{-1}/1 = 0.32$

$\Pr\{$Exactly one cell in a square$\} = 1 * e^{-1}/1 = 0.32$

$\Pr\{$Two or more cells in a square$\} = 1 - 0.32 - 0.32 = 0.36.$

Two cells was one too many. A clone or colony must begin with a single cell (which subsequently divides several times). I had to dilute the mixture a third time to ensure the percentage of squares that included two or more cells was vanishingly small. Alas, the vast majority of squares were now empty; I was forced to spend hundreds of additional hours peering through the microscope looking for the few squares that did include a clone.

3.5.2 Comparing Empirical and Theoretical Poisson Distributions

Table 3.1 compares the spatial distribution of superclusters of galaxies with a Poisson distribution that has the same expectation. The fit is not perfect due both to chance and to ambiguities in the identification of distinct superclusters.

To solve the next two exercises, make use of the R functions for the Poisson similar to those we used in studying the binomial similar including dpois, ppois, qpois, and rpois.

Exercise 3.12: Generate the results of 100 samples from a Poisson distribution with an expected number of two events per interval. Compare the graph of the resulting empirical distribution with that of the corresponding theoretical distribution. Determine the 10th, 50th, and 90th percentiles of the theoretical Poisson distribution.

Table 3.1 Spatial Distribution of Superclusters of Galaxies

Number	0	1	2	3	4	5
Observed	7178	2012	370	38	1	1
Poisson distribution	7095	2144	324	32	2	0

Exercise 3.13: Show that if $\Pr\{X = k\} = \lambda^k e^{-\lambda}/k!$ for $k = 0, 1, 2, \ldots$ that is, if X is a Poisson variable, then the expected value of $X = \Sigma_k k \Pr\{X = k\} = \lambda$.

*Exercise 3.14: Show that the variance of the Poisson distribution is also λ.

Exercise 3.15: In subsequent chapters, we'll learn how the statistical analysis of trials of a new vaccine is often simplified by assuming that the number of infected individuals follows a Poisson rather than a Binomial distribution. To see how accurate an approximation this might be, compare the cumulative distribution functions of a binomial variable, B(100,0.01) and a Poisson variable, P(1) over the range 0 to 100.

3.5.3 Comparing Two Poisson Processes

Recently, I had the opportunity to participate in the conduct of a very large-scale clinical study of a new vaccine. I'd not been part of the design team, and when I read over the protocol, I was stunned to learn that the design called for inoculating and examining 100,000 patients! 50,000 with the experimental vaccine, and 50,000 controls with a harmless saline solution.

Why so many? The disease at which the vaccine was aimed was relatively rare. In essence, we would be comparing two *Poisson* distributions. Suppose we could expect 0.8% or 400 of the controls to contract the disease, and 0.7% or 350 of those vaccinated to contract it. Put another way, if the vaccine was effective, we would expect 400/750th of the patients who contracted the disease to be controls. While if the vaccine was ineffective (no better and no worse than no vaccine at all), we would expect 50% of the patients who contracted the disease to be controls. The problem boils down to a single binomial.

Exercise 3.16: In actuality, only 200 of the controls contracted the disease along with 180 of those who had been vaccinated. What is the probability that such a difference might have occurred by chance alone?

3.6 CONTINUOUS DISTRIBUTIONS

The vast majority of the observations we make are on a continuous scale even if, in practice, we only can make them in discrete increments. For example, a man's height might actually be 1.835421117 m, but we are not likely to record a value with that degree of accuracy (nor want to). If one's measuring stick is accurate to the nearest millimeter, then the probability that an individual selected at random will be exactly 2 m tall is really the probability that his or her height will lie between 1.9995 and 2.0004 m. In such a situation, it is more convenient to replace the sum of an arbitrarily small number of quite small probabilities with an integral

$$\int_{1.9995}^{2.0004} dF[x] = \int_{1.9995}^{2.0004} f[x]dx,$$

where $F[x]$ is the cumulative distribution function of the continuous variable representing height, and $f[x]$ is its probability density. Note that $F[x]$ is now defined as

$$\int_{-\infty}^{x} f[y]dy. *$$

As with discrete variables, the cumulative distribution function is monotone nondecreasing from 0 to 1, the distinction being that it is a smooth curve rather than a step function.

The mathematical expectation of a continuous variable is

$$\int_{-\infty}^{\infty} yf[y]dy,$$

and its variance is

$$\int_{-\infty}^{\infty} (y-EY)^2 f[y]dy.$$

3.6.1   The Exponential Distribution

The simplest way to obtain continuously distributed random observations is via the same process that gave rise to the Poisson. Recall that a Poisson process is such that events in nonoverlapping intervals are independent and identically distributed. The times[†] between Poisson events follow an exponential distribution:

$$F[t|\lambda] = \Pr\{T \le t|\lambda\} = 1 - \exp[-t/\lambda] \text{ for positive values of both } t \text{ and } \lambda.$$

When t is zero, $\exp[-t/\lambda]$ is 1 and $F[t|\lambda]$ is 0. As t increases, $\exp[-t/\lambda]$ decreases rapidly toward zero and $F[t|\lambda]$ increases rapidly to 1. The rate of increase is inversely proportional to the magnitude of the parameter λ. In fact, $\log(1 - F[t|\lambda]) = -t/\lambda$. The next exercise allows you to demonstrate this for yourself.

> **Exercise 3.17:** Draw the cumulative distribution function of an exponentially distributed observation with parameter $\lambda = 3$. What is the median? Use the following R code:# using R's seq() command, divide the interval 0, 12 into increments 1/10th of a unit apart
>
> ➢ t=seq(0,12,0.1)
> ➢ F= 1 - exp(-t/3)
> ➢ plot (t,F)

---

* If it's been a long while or never since you had calculus, note that the differential dx or dy is a meaningless index, so any letter will do, just as $\Sigma_k f_k$ means exactly the same thing as $\Sigma_j f_j$.
† Time is almost but not quite continuous. Modern cosmologists now believe that both time and space are discrete, with time determined only to the nearest $10^{-23\text{rd}}$ of a second, that's 1/100000000000000000000000th of a second.

*Exercise 3.18: (requires calculus) What is the expected value of an exponentially distributed observation with parameter λ?

The times between counts on a Geiger counter follow an exponential distribution. So do the times between failures of manufactured items, like light bulbs, that rely on a single crucial component.

Exercise 3.19: When you walk into a room, you discover the light in a lamp is burning. Assuming the life of its bulb is exponentially distributed with an expectation of 1 year, how long do you expect it to be before the bulb burns out? (Many people find they get two contradictory answers to this question. If you are one of them, see W. Feller, *An Introduction to Probability Theory and Its Applications*, 1966, vol. 1, pp. 11–12.)

Most real-life systems (including that complex system known as a human being) have built-in redundancies. Failure can only occur after a series of n breakdowns. If these breakdowns are independent and exponentially distributed, all with the same parameter λ, the probability of failure of the total system at time $t > 0$ is

$$f(t) = \lambda \exp(-\lambda t)(\lambda t)^n / n!.$$

3.7 SUMMARY AND REVIEW

In this chapter, we considered the form of three common distributions, two discrete—the binomial and the Poisson, and one continuous—the exponential. You were provided with the R functions necessary to generate random samples, and probability values from the various distributions.

Exercise 3.20: Make a list of all the italicized terms in this chapter. Provide a definition for each one along with an example.

Exercise 3.21: A farmer was scattering seeds in a field so they would be at least a foot apart 90% of the time. On the average, how many seeds should he sow per square foot?

The answer to Exercise 3.1 is yes, of course; an observation or even a sample of observations from one population may be larger than observations from another population even if the vast majority of observations are quite the reverse. This variation from observation to observation is why before a drug is approved for marketing, its effects must be demonstrated in a large number of individuals and not just in one or two. By the same token, when reporting the results of a survey, we should always include the actual frequencies and not just percentages.

Chapter 4

Estimation and the Normal Distribution

In this chapter, we introduce and make use of a probability distribution that often arises in practice, albeit indirectly, the "normal" or "Gaussian" distribution. An observation that is the sum of a large number of factors, each of which makes only a small contribution to the total, will have a normal distribution. Thus, the mean of a large number of independent observations on the elements of a homogeneous population will have a normal distribution.

You also will learn in this chapter about the desirable properties of both point and interval estimates and how to apply this knowledge to estimate the parameters of a normal distribution. You'll be provided with the R functions you need test hypotheses about the parameters.

4.1 POINT ESTIMATES

We are often required to estimate some property of a population based on a random representative sample from that population. For example, we might want to estimate the population mean or the population variance.

A desirable estimator will be *consistent*, that is, as the sample grows larger, an estimate based on the sample will get closer and closer to the true value of the population parameter. The sample mean is a consistent estimator of the population mean. The sample variance is a consistent estimator of the population variance.

When we make decisions based on an estimate h of a population parameter θ, we may be subject to *losses* based on some function L of the difference between our estimate and the true value, $L(h, \theta)$. Suppose this loss function is proportional to the square of the difference, $L(h, \theta) = k(h - \theta)^2$, as in Figure 4.1.

Then we would want to make use of an estimate that minimizes the *mean squared error*, that is, an estimate that minimizes the expected value of $(h - \theta)^2$, which is equal to the variance of the statistic h plus $(Eh - \theta)^2$, where Eh denotes the expected value of the estimate when averaged over a large number of samples. The

Introduction to Statistics Through Resampling Methods and R, Second Edition. Phillip I. Good.
© 2013 John Wiley & Sons, Inc. Published 2013 by John Wiley & Sons, Inc.

59

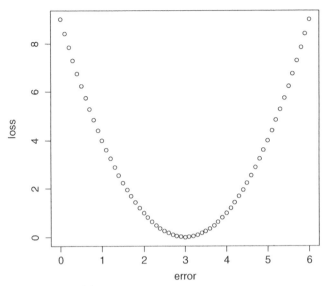

Figure 4.1 Loss as a result of an error in estimation.

first term of this sum, the variance of the statistic h, is a measure of the *precision* of our estimate, the second term $(Eh - \theta)^2$, called the *bias*, is a measure of its *accuracy*. An *unbiased* estimate h of θ is such that $Eh = \theta$.

The sample mean and the sample variance are both consistent, unbiased, minimum variance, minimum squared-error estimators of the population mean and the population variance respectively.

Statistics and Parameters

A *statistic* is any single value, such as the sample mean

$$\frac{1}{n}\sum_{k=1}^{n} X_k,$$

that is a function of the observations. Its value varies from sample to sample. A parameter is any single value, such as the mean θ of a normal distribution, that summarizes some aspect of an entire population. Its value is fixed.

Examples of sample statistics include measures of location and central tendency, such as the sample mode, sample median, and sample mean, extrema, such as the sample minimum and maximum, and measures of variation, and dispersion, such as the sample standard deviation. These same measures are considered *parameters* when they refer to an entire population, for example, the population mean, population range, and population variance.

In subsequent chapters, we will use sample statistics to estimate the values of population parameters and to test hypotheses about them.

4.2   PROPERTIES OF THE NORMAL DISTRIBUTION

The properties of the normal distribution—its percentiles, for example, are completely determined by its mean and variance. Figures 4.2 and 4.3 depict the cumulative distribution function and the probability density curve of a normally distributed population with zero mean and unit variance.

To graph a normal probability distribution with some other standard deviation, simply interpret the scale on the x-axis of these figures as reading in units of standard deviations. To graph a normal probability distribution with some other mean, simply shift the scale on the x-axis so the value of the new mean is located where the zero used to be.

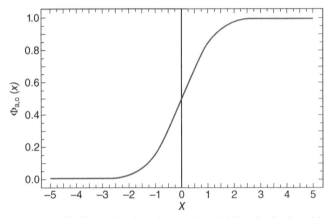

Figure 4.2   Cumulative distribution function of a normal probability distribution with zero mean and unit variance.

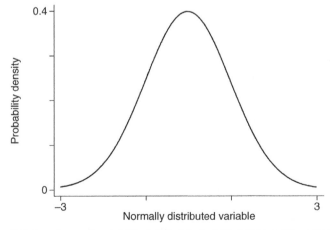

Figure 4.3   Bell-shaped normal probability density. Interpret the density on the y-axis as the relative probability that an observation from a normal distribution with zero mean and unit variance will take the value shown on the x-axis.

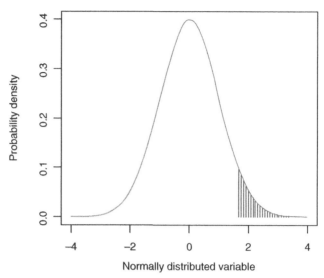

Figure 4.4   The ratio of the shaded area to the total area under the curve yields the probability that a normally distributed random variable will be at least 1.64 standard deviations above its mean.

The probability densities depicted in Figure 4.3 should be interpreted as relative rather than absolute probabilities. The normal distribution is that of a continuous random variable, so that the probability of observing any specific value, say 1.000000000, is zero. But we can and will talk about the probability that a normally distributed variable will lie inside some fixed interval. We can obtain this value by calculating the area under the density curve. For example, the shaded area in Figure 4.4 reveals that the probability is 10% that a normally distributed random variable will be greater than 1.64 standard deviations larger than the mean. To be honest, I didn't read this value from the graph, but let R determine the probability for me, with the function pnorm(1.64).

Exercise 4.1: The heights of Cambridge students in the early 1900s were found to follow a normal distribution. The heights of today's students do not. Why do you think this is?

To generate a plot similar to Figure 3.3 of the empirical cumulative probability distributions of samples of various sizes taken from a normal probability distribution with mean zero and variance 1, I used the following R code:

```
➤ F1600=ecdf(rnorm(1600,0,1))
➤ plot(F1600)
➤ F100=ecdf(rnorm(100))
➤ lines(F100,pch=".")
➤ F10=ecdf(rnorm(100))
➤ lines(F10, verticals= TRUE, do.points = FALSE)
```

Exercise 4.2: Draw the empirical cumulative probability density function of a sample of size 1000 from a normal distribution with mean 2.5 and variance 9.

To see why the normal distribution plays such an important role in statistics, please complete Exercise 4.3, which requires you to compute the distribution of the mean of a number of binomial observations. As you increase the number of observations used to compute the mean from 5 to 12 so that each individual observation makes a smaller relative contribution to the total, you will see that the distribution of means looks less and like the binomial distribution from which they were taken and more and more like a normal distribution. This result will prove to be true regardless of the distribution from which the observations used to compute the mean are taken, providing that this distribution has a finite mean and variance.

Exercise 4.3: Generate five binomial observations based on 10 trials with probability of success p = 0.35 per trial. Compute the mean value of these five. Repeat this procedure 512 times, computing the mean value each time. Plot the empirical cumulative distribution function of these means. Compare with the empirical cumulative distribution functions of (1) a sample of 512 normally distributed observations with expected value 3.5 and variance 2.3, and (2) a sample of 512 binomial observations, each consisting of 10 trials with a probability of success of $p = 0.4$ per trial. Repeat the entire exercise, computing the mean of 12 rather than 5 binomial observations.

Here is the R code you'll need to do this exercise:

```
#Program computes the means of N samples of k
  #binomial(10,0.35) variables
#set number of times to compute a mean
N = 512
#set number of observations in sample
k = 5
#create a vector in which to store the means
stat = numeric(N)
#Set up a loop to generate the N means
for (i in 1:N) stat[i] = mean(rbinom (k,10, 0.35))
plot(ecdf(stat,14))

#generate a sample of N normally distributed
  #observations
sampN = rnorm (N, 3.5, 2.3)
plot(ecdf (sampN,14))

#generate a sample of N binomial observations
sampB = rbinom (N, 10, 0.35)
plot(ecdf (sampB,14))
```

4.2.1 Student's *t*-Distribution

The probability is 90% that a normally distributed random variable drawn at random will lie within 1.65 standard deviations of its mean. If we knew the standard deviation of the population from which our mean was drawn, we could use this knowledge to obtain an *interval estimate* of the unknown population mean based on the sample mean. But we don't; we only know the sample standard deviation.

Fortunately, we can make use of the statistic

$$t = \frac{\bar{X} - \theta}{s / \sqrt{n}},$$

which has a Student's t-distribution with $n - 1$ degrees of freedom, where n is the sample size, θ is the population mean and s is the standard deviation of the sample.*

The ratio of the standard deviation divided by the square root of the sample size is often referred to as the *standard error of the mean.*

Suppose we have the following sample,

```
> data = rnorm(8)
[1]  0.8011477  0.4213278 -0.4574428  1.8352498
    -0.5901639 -0.8767256  1.0373933
[8] -0.4187433
```

The R function t.test will provide us with a 90% confidence interval for the mean.

```
> t.test(data,conf.level=.9)
  One Sample t-test
data: data
t = 0.6488, df = 7, p-value = 0.5372
alternative hypothesis: true mean is not equal to 0
90 percent confidence interval: -0.4205535  0.8585642
sample estimates: mean of x 0.2190054
```

The probability is 90% that the confidence interval (–0.42, 0.86) covers the true unknown value of the mean.

Note that we have discarded the least significant digits when we reported this result.

Exercise 4.4: Find a 90% confidence interval for the mean based on the following sample, 5.1451036 –1.5395632 1.8893192 –2.1733570 –0.9131787 –0.5619511 0.8190254.

Exercise 4.5: Will an 80% confidence interval for the mean be longer or shorter than a 90% confidence interval? Why?

4.2.2 Mixtures of Normal Distributions

Many real-life distributions strongly resemble mixtures of normal distributions. Figure 4.5 depicts just such an example in the heights of California sixth graders. Though the heights of boys and girls overlap, it is easy to see that the population of sixth graders is composed of a mixture of the two sexes.

If you suspect you are dealing with a mixture of populations, it is preferable to analyze each subpopulation separately. We'll discuss more about this approach in Chapter 7 and Chapter 8.

---

* That missing degree of freedom came about from our need to use an estimate of the variance based on the sample.

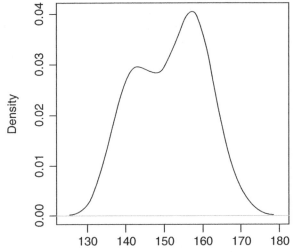

Figure 4.5   Distribution of the heights of California sixth graders.

Exercise 4.6: Provide an 80% confidence interval for the mean height of sixth-grade students using the data of Chapter 1.

4.3 USING CONFIDENCE INTERVALS TO TEST HYPOTHESES

A fast-food restaurant claims that $750 of their daily revenue is from the drive-thru window. They've collected 2 weeks of receipts from the restaurant and turned them over to you. Each day's receipt shows the total revenue and the "drive-thru" revenue for that day.

They do not claim their drive-thru produces $750 of their revenue, day in and day out, only that their over-all average is $750.

Drive-Thru Sales: 754 708 652 783 682 778 665 799 693 825 828 674 792 723

Exercise 4.7: Derive a 90% confidence interval for the mean daily revenue via drive-thru. If $750 is included in this interval, then accept the hypothesis.

4.3.1 Should We Have Used the Bootstrap?

We've already made use of the percentile or uncorrected bootstrap on several occasions, first to estimate precision and then to obtain interval estimates for population parameters. Readily computed, the bootstrap seems ideal for use with the drive-thru problem. Still, if something seems too good to be true, it probably is. Unless corrected, bootstrap interval estimates are *inaccurate* (i.e., they will include the true value of the unknown parameter less often than the stated confidence probability) and *imprecise* (i.e., they will include more erroneous values of the unknown parameter than is desirable). When the original samples contain less than a hundred observations, the confidence bounds based on the primitive bootstrap may vary widely from simulation to simulation.

What this means to us is that even if the figure 75% lies outside a 90% bootstrap confidence interval, we nay run the risk of making an error more than 100% − 90% = 10% of the time in rejecting this hypothesis.

4.3.2   The Bias-Corrected and Accelerated Nonparametric Bootstrap

The percentile bootstrap is most accurate when the observations are drawn from a normal distribution. The bias-corrected-and-accelerated bootstrap takes advantage of this as follows: Suppose θ is the parameter we are trying to estimate, $\hat{\theta}$ is the estimate, and we are able to come up with a transformation m such that $m(\theta)$ is normally distributed about $m(\hat{\theta})$. We could use this normal distribution and the bootstrap to obtain an unbiased confidence interval, and then apply a back-transformation to obtain an almost-unbiased confidence interval.

The good news is that we don't actually need to determine what the function m should be or compute it to obtain the desired interval. To use R for this purpose, all you need is to download and install the "Boot" package of functions and make use of the functions **boot()** and **boot.ci()**.

Make sure you are connected to the Internet and then pull down on the "Packages" menu as shown in Figure 4.6 and click on "Install package(s)…"

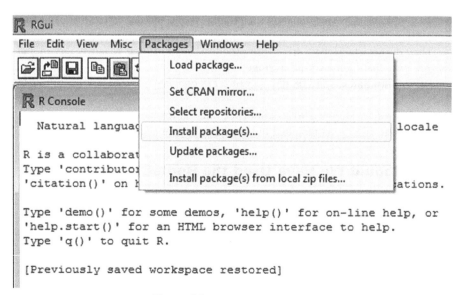

Figure 4.6   R's packages menu.

Figure 4.7   (a) List of mirror sites that can be used for downloading R packages. (b) List of R packages.

Select the nearest CRAN mirror as shown in Figure 4.7a and then the Package "boot" as shown in Figure 4.7b.

The installation, which includes downloading, unzipping, and integrating the new routines, is done automatically. The installation needs to be done once and once only. But each time before you can use any of the boot library routines, you'll need to load the supporting functions into computer memory by typing

```
> library(boot)
```

R imposes this requirement to keep the amount of memory its program uses to the absolute minimum.

We'll need to employ two functions from the boot library.

The first of these functions **boot (Data, Rfunction, number)** has three principal arguments. **Data** is the name of the data set you want to analyze, **number** is the number of bootstrap samples you wish to draw, and **Rfunction** is the name of an R function you must construct separately to generate and hold the values of existing R statistics functions, such as **median()** or **var()**, whose value you want a bootstrap interval estimate of. For example,

```
> f.median<- function(y,id){
+ median( y[id])
+ }
```

where R knows id will be a vector. Then

```
> boot.ci(boot(classdata, f.median, 400), conf = 0.90)
```

will calculate a 90% confidence interval for the median of the classdata based on 400 simulations.

4.3.3 The Parametric Bootstrap

When observations are drawn from a very skewed (and thus very unlike a normal) distribution, such as the exponential, the parametric bootstrap can prove of value in obtaining confidence intervals for the population parameters.

It's a two-step method:

1. We draw a bootstrap sample
2. We use the sample to obtain point estimates of the parameters of the distribution we assume the original sample was drawn from.

This parametric approach is of particular value when we are trying to estimate one of the tail percentiles such as P_{10} or P_{90}, for the sample alone seldom has sufficient information.

For example, if we know this population has an exponential distribution, we would use the sample mean to estimate the population mean. Then, we would draw a series of random samples of the same size as our original sample from an exponential distribution whose mathematical expectation was equal to the sample mean to obtain a confidence interval for the population parameter of interest.

The following R program uses an exponential distribution as the basis of a parametric bootstrap to obtain a 90% confidence interval for the interquartile range (IQR) of the population from which the data set A was taken. If P_{75} and P_{25} are the 75th and 25th percentiles of the population, IQR is defined as $P_{75} - P_{25}$. This program makes use of the results of Exercise 3.16, which tells us that the expected value of an exponential distribution with parameter λ is $1/\lambda$.

```
> N = 1000 #number of bootstrap samples
> #create a vector in which to store the IQR's
> IQR = numeric(N)
> #Set up a loop to generate the IQR's
> for (i in 1:N) {
```

```
    bA=sample (A, n, replace = T)
    IQR[i] = qexp(.75,1/mean(bA))-qexp(.25, 1/mean(bA))
    }
➢ quantile (IQR , probs = c(.05,.95))
```

Exercise 4.8: Given the following data in years drawn from a distribution of failure times of an electronic component, find a confidence interval for the data's median: 1.071 4.397 2.733 0.484 0.323 1.158 1.962 0.660 0.070 0.946 0.212 3.038

4.4 PROPERTIES OF INDEPENDENT OBSERVATIONS

As virtually all the statistical procedures in this text require that our observations be independent of one another, a study of the properties of independent observations will prove rewarding. Recall from the previous chapter that if X and Y are independent observations, then

$$\Pr\{X = k \text{ and } Y = j\} = \Pr\{X = k\}\Pr(Y = j\},$$

and

$$\Pr\{X = k|Y = j\} = \Pr\{X = k\}.$$

If X and Y are independent discrete random variables and their expectations exist and are finite,* then $E\{X + Y\} = EX + EY$. To see this, suppose that $Y = y$. The conditional expectation of $X + Y$ given $Y = y$ is

$$\begin{aligned}
E\{X + Y|Y = y\} &= \Sigma_k (k + y)\Pr\{X + Y = k + y|Y = y\} \\
&= \Sigma_k (k + y)\Pr\{X = k\} \\
&= EX + y\Sigma_k \Pr\{X = k\} = EX + y.
\end{aligned}$$

Taking the average of this conditional expectation over all possible values of Y, yields

$$E\{X + Y\} = \Sigma_j (EX + j)\Pr\{Y = j\} = EX*1 + EY.$$

A similar result holds if X and Y have continuous distributions, providing that their individual expectations exist and are finite.

Exercise 4.9: Show that for any variable X with a finite expectation, $E(aX) = aEX$, where a is a constant. Here is one example: The mean of heights measured in inches will be 12 times their mean when measured in feet.

Exercise 4.10: Show that the expectation of the mean of n independent identically distributed random variables with finite expectation θ is also θ.

---

* In real life, expectations almost always exist and are finite—the expectations of ratios are a notable exception.

The Cauchy Distribution: An Exception to the Rule

It seems obvious that every distribution should have a mean and a variance. But telling a mathematician something is "obvious" only encourages them to find an exception. The Cauchy distribution is just one example: The expression $f(x) = 1/\pi(1 + x)$ for $-\infty < x < \infty$ is a probability density since $\int f(x)dx = 1$, but neither its mean nor its variance exists. Does the Cauchy distribution arise in practice? It might if you study the ratio X/Y of two independent random variables each distributed as $N(0,1)$.

The variance of the sum of two independent variables X and Y is the sum of their variances, providing each of these variances exists and is finite.

Exercise 4.11: Given the preceding result, show that the variance of the mean of n independent identically distributed observations is $1/n$th of the variance of just one of them. Does this mean that the arithmetic mean is more precise than any individual observation? Does this mean that the sample mean will be closer to the mean of the population from which it is drawn than any individual observation would be, that is, that it would be more accurate?

*Exercise 4.12: Show that the sun of n independent identically distributed normal distributions with mean θ (theta) and variance σ^2 (sigma squared) has a normal distribution with mean $n\theta$ and variance $n\sigma^2$.

4.5   SUMMARY AND REVIEW

In this chapter, you leaned to distinguish statistics, functions of the observations, and the estimates based upon them, both of which vary from sample to sample, from parameters, fixed nonvarying property of a population.

You considered the form and properties of one of the most common probability distributions—the normal. The distribution of the sample mean of most distributions encountered in practice will take on the form of a normal distribution even with moderate-size samples.

You learned that the sample mean and sample variance are desirable point estimates of the equivalent population parameters because they are both accurate and precise. You learned how the Student's t-distribution might be used to obtain confidence intervals for the mean of a distribution of near-normal measurements.

You made use of the R functions rnorm(), qexp(), ecdf(), plot(), and lines(). You were shown how to download and use libraries to enlarge R's capabilities: library (boot), for example, has the function boot.ci()with which to perform the bias-corrected and accelerated bootstrap (BCa) bootstrap. You learned how to use either the BCa or the parametric bootstrap to obtain confidence intervals of conditional distributions for properties of the distribution other than the mean.

We expanded on the properties of independent observations, properties we will make frequent use of in future chapters.

Exercise 4.13: Make a list of all the italicized terms in this chapter. Provide a definition for each one along with an example.

Chapter 5

Testing Hypotheses

In this chapter, we develop improved methods for testing hypotheses by means of the bootstrap, introduce permutation testing methods, and apply these and the *t*-test to experiments involving one and two samples. We then address the obvious but essential question: How do we choose the method and the statistic that is best for the problem at hand?

5.1 TESTING A HYPOTHESIS

Suppose we were to pot a half-dozen tomato plants in ordinary soil and a second half-dozen plants in soil enriched with fertilizer. If we wait a few months, we can determine whether the addition of fertilizer increases the resulting yield of tomatoes, at least as far as these dozen plants are concerned. But can we extend our findings to all tomatoes? Or, at least, to all tomatoes of the same species that are raised under identical conditions apart from the use of fertilizer.

To ensure that we can extend our findings, we need to proceed as follows: First, the 12 tomato plants used in our study must be a *random sample* from a nursery. If we choose only plants with especially green leaves for our sample, then our results can be extended only to plants with especially green leaves. Second, we have to divide the 12 plants into two treatment groups at random. If we subdivide by any other method, such as tall plants in one group and short plants in another, then the experiment would not be about fertilizer but about our choices.

I performed just such a randomized experiment a decade or so ago. Only, I was interested in the effects vitamin E might have on the aging of human cells in culture. After several months of abject failure—contaminated cultures, spilled containers—I succeeded in cloning human diploid fibroblasts in eight culture dishes. Four of these dishes were filled with a conventional nutrient solution and four held a potential "life-extending" solution to which vitamin E had been added. All the cells in the dishes came from the same culture so that the initial distribution of cells was completely random.

I waited 3 weeks with my fingers crossed—there is always a risk of contamination with cell cultures—but at the end of this test period, three dishes of each type

Introduction to Statistics Through Resampling Methods and R, Second Edition. Phillip I. Good.
© 2013 John Wiley & Sons, Inc. Published 2013 by John Wiley & Sons, Inc.

had survived. I transplanted the cells, let them grow for 24 hours in contact with a radioactive label, and then fixed and stained them before covering them with a photographic emulsion.

Ten days passed and we were ready to examine the autoradiographs. "121, 118, 110, 34, 12, 22." I read and reread these six numbers over and over again. The larger numbers were indicative of more cell generations and an extended lifespan. If the first three generation-counts were from treated colonies and the last three were from untreated, then I had found the fountain of youth. Otherwise, I really had nothing to report.

5.1.1 Analyzing the Experiment

How had I reached this conclusion? Let's take a second, more searching look. First, we identify the primary hypothesis and the alternative hypothesis of interest.

I wanted to assess the life-extending properties of a new experimental treatment with vitamin E. To do this, I had divided my cell cultures at random into two groups: one grown in a standard medium and one grown in a medium containing vitamin E. At the conclusion of the experiment and after the elimination of several contaminated cultures, both groups consisted of three independently treated dishes.

My primary hypothesis was a *null hypothesis*, that the growth potential of a culture would not be affected by the presence of vitamin E in the media: All the cultures would have equal growth potential. The *alternative* hypothesis of interest was that cells grown in the presence of vitamin E would be capable of many more cell divisions. This is a *one-sided* alternative hypothesis.

Under the null hypothesis, the labels "treated" and "untreated" provide no information about the outcomes: The observations are expected to have more or less the same values in each of the two experimental groups. If they do differ, it should only be as a result of some uncontrollable random fluctuations. Thus, if this null or no-difference-among-treatments hypothesis were true, I was free to exchange the treatment labels.

The alternative is a distributional shift like that depicted in Figure 5.1 in which greater numbers of cell generations are to be expected as the result of treatment with vitamin E (though the occasional smaller value cannot be ruled out completely).

The next step is to choose a test statistic that discriminates between the hypothesis and the alternative. The statistic I chose was the sum of the counts in the group treated with vitamin E. **If** the alternative hypothesis is true, most of the time this sum ought to be larger than the sum of the counts in the untreated group. *If the null hypothesis* is true, that is, if it doesn't make any difference which treatment the cells receive, then the sums of the two groups of observations should be approximately the same. One sum might be smaller or larger than the other by chance, but most of the time the two shouldn't be all that different.

The third step is to compute the test statistic for each of the possible reassignments of treatment label to result and compare these values with the value of the test statistic as the data were labeled originally. As it happened, the first three

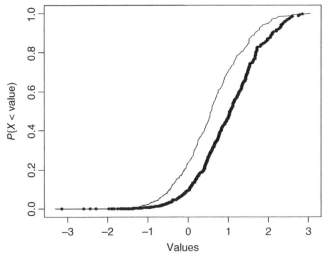

Figure 5.1   The shift to the right of the darker curve represents the potential life-extending effect of treatment with vitamin E.

observations—121, 118, and 110—were those belonging to the cultures that received vitamin E. The value of the test statistic for the observations as originally labeled is $349 = 121 + 118 + 110$.

I began to rearrange (permute) the observations, by randomly reassigning the six labels, three "treated" and three "untreated," to the six observations. For example: treated, 121 118 34, and untreated, 110 12 22. In this particular rearrangement, the sum of the observations in the first (treated) group is 273. I repeated this step until all $\binom{6}{3} = 20$ distinct rearrangements had been examined.*

	First Group	Second Group	First Group
		Sum of	
1.	121 118 110	34 22 12	349
2.	121 118 34	110 22 12	273
3.	121 110 34	118 22 12	265
4.	118 110 34	121 22 12	262
5.	121 118 22	110 34 12	261
6.	121 110 22	118 34 12	253
7.	121 118 12	110 34 22	251
			(*Continued*)

---

* Determination of the number of relabelings, "6 choose 3" in the present case, is considered in Section 2.2.1.

	First Group	Second Group	First Group
		Sum of	
8.	118 110 22	121 34 12	250
9.	121 110 12	118 34 22	243
10.	118 110 12	121 34 22	240
11.	121 34 22	118 110 12	177
12.	118 34 22	121 110 12	174
13.	121 34 12	118 110 22	167
14.	110 34 22	121 118 12	166
15.	118 34 12	121 110 22	164
16.	110 34 12	121 118 22	156
17.	121 22 12	118 110 34	155
18.	118 22 12	121 110 34	152
19.	110 22 12	121 118 34	144
20.	34 22 12	121 118 110	68

The sum of the observations in the original vitamin E-treated group, 349, is equaled only once and never exceeded in the 20 distinct random relabelings. If chance alone is operating, then such an extreme value is a rare, only a 1-time-in-20 event. If I reject the null hypothesis and embrace the alternative that the treatment is effective and responsible for the observed difference, I only risk making an error and rejecting a true hypothesis once in every 20 times.

In this instance, I did make just such an error. I was never able to replicate the observed life-promoting properties of vitamin E in other repetitions of this experiment. Good statistical methods can reduce and contain the probability of making a bad decision, but they cannot eliminate the possibility.

> **Exercise 5.1:** How was the analysis of the cell culture experiment affected by the loss of two of the cultures due to contamination? Suppose these cultures had escaped contamination and given rise to the observations 90 and 95; what would be the results of a permutation analysis applied to the new, enlarged data set consisting of the following cell counts
>
> ```
> Treated   121 118 110 90      Untreated  95 34 22 12?
> ```
>
> *Hint*: To determine how probable an outcome like this is by chance alone, first determine how many possible rearrangements there are. Then list all the rearrangements that are as or more extreme than this one.

5.1.2 Two Types of Errors

In the preceding example, I risked rejecting the null hypothesis in error 5% of the time. Statisticians call this making a *Type I error* and they call the 5%, the *signifi-*

Table 5.1 Decision Making Under Uncertainty

	Investigator's decision	
The facts	Not a carcinogen	Compound a carcinogen
Not a carcinogen		*Type I error*: Manufacturer misses opportunity for profit. Public denied access to effective treatment.
Carcinogen	*Type II error*: Manufacturer sued. Patients die; families suffer.	

cance level. In fact, I did make such an error, for in future experiments, vitamin E proved to be valueless in extending the life span of human cells in culture.

On the other hand, suppose the null hypothesis had been false, that treatment with vitamin E really did extend life span, and I had failed to reject the null hypothesis. Statisticians call this making a *Type II error*.

The consequences of each type of error are quite different and depend upon the context of the investigation. Consider the table of possibilities (Table 5.1) arising from an investigation of the possible carcinogenicity of a new headache cure.

We may luck out in that our samples supports the correct hypothesis, but we always run the risk of making either a Type I or a Type II error. We can't avoid it. If we use a smaller significance level, say 1%, then, if the null hypothesis is false, we are more likely to make a Type II error. If we always accept the null hypothesis, a significance level of 0%, then we guarantee we'll make a Type II error if the null hypothesis is false. This seems kind of a stupid: why bother to gather data if you're not going to use the data? But if you read or live Dilbert, then you know this happens all the time.

Exercise 5.2: The nurses have petitioned the CEO of a hospital to allow them to work 12-hour shifts. He wants to please them but is afraid that the frequency of errors may increase as a result of the longer shifts. He decides to conduct a study and to test the null hypothesis that there is no increase in error rate as a result of working longer shifts against the alternative that the frequency of errors increases by at least 30%. Describe the losses associated with Type I and Type II errors.

Exercise 5.3: Design a study. Describe a primary hypothesis of your own along with one or more likely alternatives. The truth or falsity of your chosen hypothesis should have measurable monetary consequences. If you were to test your hypothesis, what would be the consequences of making a Type I error? A Type II error?

Exercise 5.4: Suppose I'm (almost) confident that my candidate will get 60% or more of the votes in the next primary. The alternative that scares me is that she will get 45% or less. To see if my confidence is warranted, I decide to interview 20 people selected at random in a shopping mall and reject my hypothesis if 7 or fewer say they will vote for her.

a. How will the results of my interviews be distributed?
b. What is the probability of my making a Type I error?
c. What is the probability of my retaining confidence in my candidate if only 45% of the general population favor her, that is, committing a Type II error?
d. How can I reduce the probability of making a Type II error while keeping the probability of making a Type I error the same?

Exercise 5.5: Individuals were asked to complete an extensive questionnaire concerning their political views and eating preferences. Analyzing the results, a sociologist performed 20 different tests of hypotheses. Unknown to the sociologist, the null hypothesis was true in all 20 cases. What is the probability that the sociologist rejected at least one of the hypotheses at the 5% significance level? (Reread Chapter 2 and Chapter 3 if you are having trouble with this exercise.)

5.2 ESTIMATING EFFECT SIZE

In the previous example, we developed a test of the null hypothesis of no treatment effect against the alternative hypothesis that a positive effect existed. But in many situations, we would also want to know the magnitude of the effect. Does vitamin E extend cell lifespan by 3 cell generations? By 10? By 15?

In Section 1.6.2, we used the bootstrap to obtain estimates of the precision of an estimate of the sample mean or median (or, indeed, any sample percentile). In Section 1.6.2, we showed how to use the bootstrap estimate the precision of the sample mean or median (or, indeed, almost any sample statistic) as an estimator of a population parameter. As a byproduct, we obtain an *interval estimate* of the corresponding population parameter.

For example, if P_{05} and P_{95} are the 5th and 95th percentiles of the bootstrap distribution of the median of the law school LSAT data you used for Exercise 1.13, then the set of values between P_{05} and P_{95} provides a 90% *confidence interval* for the median of the population from which the data were taken.

Exercise 5.6: Obtain an 85% confidence interval for the median of the population from which the LSAT data were taken.

Exercise 5.7: Can this same bootstrap technique be used to obtain a confidence interval for the 90th percentile of the population? For the maximum value in the population?

5.2.1 Effect Size and Correlation

Suppose we have collected random samples from two populations and want to know the following:

- Do the populations from which they are drawn have the same means?
- If the means are not the same, then what is the difference between them?

To find out, we would let each sample stand in place of the population from which it is drawn, take a series of bootstrap samples separately from each sample, and compute the difference in means each time.

Suppose our data is stored in two vectors called "control" and "treated." The R code to obtain the desired interval estimate would be as follows:

```
> #This program selects 400 bootstrap samples from
  #your data
> #and then produces an interval estimate of the
  #difference in population means
> #Record group sizes
> n = length(control)
> m = length(treated)
> #set number of bootstrap samples
> N = 400
> stat = numeric(N)
  #create a vector in which to store the results
> #the elements of this vector will be
  #numbered from 1 to N
> #Set up a loop to generate a series of bootstrap
  #samples
> for (i in 1:N){
>       #bootstrap sample counterparts are denoted with a "B"
>           controlB = sample (control, n, replace = T)
>           treatB = sample (treated, m, replace = T)
>           stat[i] = mean(treatB) - mean(controlB)
>           }
> quantile (x = stat, probs = c(.25,.75))
```

Yet another example of the bootstrap's application lies in the measurement of the *correlation* or degree of agreement between two variables. The Pearson correlation of two variables X and Y is defined as the ratio of the covariance between X and Y and the product of the standard deviations of X and Y. The covariance of X and Y is given by the formula

$$\sum_{k=1}^{n} (X_k - \bar{X})(Y_k - \bar{Y})/(n-1).$$

The reason we divide by the product of the standard deviations in assessing the degree of agreement between two variables is that it renders the correlation coefficient free of the units of measurement.

Recall that if X and Y are independent, $E(XY) = (EX)(EY)$, so that the expected value of the covariance and hence the correlation of X and Y is zero. If X and Y increase more or less together as do, for example, the height and weight of individuals, their covariance and their correlation will be positive so that we say that height and weight are positively correlated. I had a boss—more than one, who believed that the more abuse and criticism he heaped on an individual, the more work he could get out of them. Not. Abuse and productivity are negatively correlated; heap on the abuse and work output declines.

If $X = -Y$, so that the two variables are totally dependent, the correlation coefficient, usually represented in symbols by the Greek letter ρ (rho) will be -1. In all cases, $-1 \leq \rho \leq 1$.

> **Exercise 5.8:** Using the LSAT data from Exercise 1.13 and the bootstrap, obtain an interval estimate for the correlation between the LSAT score and the student's subsequent GPA. Compare with the result using Pearson correlation, **cor.test**().

> **Exercise 5.9:** Obtain an interval estimate for the correlation using the bias-corrected and accelerated bootstrap discussed in the previous chapter.

> **Exercise 5.10:** Trying to decide whether to take a trip to Paris or Tokyo, a Canadian student kept track of how many Euros and Yen his dollars would buy. Month by month he found that the values of both currencies were rising. Does this mean that improvements in the European economy are reflected by improvements in the Japanese economy?

5.2.2 Using Confidence Intervals to Test Hypotheses

Suppose we have derived a 90% confidence interval for some parameter, for example, a confidence interval for the difference in means between two populations, one of which was treated and one that was not. We can use this interval to test the hypothesis that the difference in means is 4 units, by accepting this hypothesis if 4 is included in the confidence interval and rejecting it otherwise. If our alternative hypothesis is nondirectional and two-sided, $\theta_A \neq \theta_B$, the test will have a Type I error of $100\% - 90\% = 10\%$.

Clearly, hypothesis tests and confidence intervals are intimately related. Suppose we test a series of hypotheses concerning a parameter θ. For example, in the vitamin E experiment, we could test the hypothesis that vitamin E has no effect, $\theta = 0$, or that vitamin E increases life span by 25 generations, $\theta = 25$, or that it increases it by 50 generations, $\theta = 50$. In each case, whenever we accept the hypothesis, the corresponding value of the parameter should be included in the confidence interval.

In this example, we are really performing a series of one-sided tests. Our hypotheses are that $\theta = 0$ against the one-sided alternative that $\theta > 0$, that $\theta \leq 25$ against the alternative that $\theta > 25$, and so forth. Our corresponding confidence interval will be one-sided also; we will conclude $\theta < \theta_U$ if we accept the hypothesis $\theta = \theta_0$ for all values of $\theta_0 < \theta_U$ and reject it for all values of $\theta_0 \geq \theta_U$.

Suppose, instead, we'd wanted to make a comparison between two treatments of unknown value or between two proposed direct-mail appeals for funds. Our null hypothesis would be that it didn't make any difference which treatment or which marketing letter was used; the two would yield equivalent results. Our *two-sided alternative* would be that one of the two would prove superior to the other. Such a two-sided alternative would lead to a two-sided test in which we would reject our null hypothesis if we observed either very large or very small values of the test statistic. Not surprisingly, one-sided tests lead to one-sided confidence intervals and two-sided tests to two-sided confidence intervals.

Exercise 5.11: What is the relationship between the significance level of a test and the confidence level of the corresponding interval estimate?

Exercise 5.12: In each of the following instances, would you use a one-sided or a two-sided test?

a. Determine whether men or women do better on math tests.
b. Test the hypothesis that women can do as well as men on math tests.
c. In Commonwealth v. Rizzo et al., 466 F. Supp 1219 (E.D. Pa 1979), help the judge decide whether certain races were discriminated against by the Philadelphia Fire Department by means of an unfair test.

Exercise 5.13: Use the data of Exercise 5.1 to derive an 80% upper confidence bound for the effect of vitamin E to the nearest five cell generations.

5.3 APPLYING THE *T*-TEST TO MEASUREMENTS

To test hypothesis concerning the means of measurements using Student's *t*, we need first answer three questions:

1. Is there sufficient data? The closer the distribution from which the observations are drawn to a normal distribution, the fewer the number of observations that are required. With too few observations, we would make use instead of the permutation tests described in Section 5.1.1 and Section 5.4.

2. Will we be analyzing one sample or two? (The analysis of three or more samples is considered in Chapter 8.)

3. Is the test one-sided or two-sided? That is, are we testing for improvement only or for any kind of difference, positive or negative?

For example, suppose we wish to test of the hypothesis that daily drive-in sales average $750, not just for our sample of sales data, but also for past and near-future sales. We have 14 observations, which is a sufficient number to apply the one-sample, two-sided *t*-test.

Entering the R commands in the R interpreter and using the data from Section 4.3

```
> t.test(sales, alternativ e="t",mu = 750)
```

yields the following output:

```
        One Sample t-test
data:  sales
t = -0.6253,  df = 13,  p-value = 0.5426
alternative hypothesis: true mean is not equal to 750
```

```
95 percent confidence interval:
704.1751 775.2535
sample estimates: mean of x 739.7143
```

Exercise 5.14: In describing the extent to which we might extrapolate from our present sample of drive-in data, we used the qualifying phrase "near-future." Is this qualification necessary or would you feel confident in extrapolating from our sample to all future sales at this particular drive-in? If not, why not?

Exercise 5.15: While some variation is be expected in the width of screws coming off an assembly line, the ideal width of this particular type of screw is 10.00, and the line should be halted if it looks as if the mean width of the screws produced will exceed 10.01 or fall below 9.99. Based on the following 10 observations, would you call for the line to halt so they can adjust the milling machine: 9.983, 10.020, 10.001, 9.981, 10.016, 9.992, 10.023, 9.985, 10.035, and 9.960?

Exercise 5.16: In the preceding exercise, what kind of economic losses do you feel would be associated with Type I and Type II errors?

5.3.1 Two-Sample Comparison

Suppose we were to teach two different groups of children: One group is taught by a conventional method; the other is taught comparable material by an entirely new approach. The following are the test results:

```
conventional 65 79 90 75 61 85 98 80 97 75
new 90 98 73 79 84 81 98 90 83 88
```

To determine whether there is a statistically significant difference between the two samples, use the following R command:

```
> t.test (new, conventional, "g")
```

Exercise 5.17: Using the above example, distinguish between statistical and practical significance.

5.3.2 Paired *t*-Test

To reduce the effects of person-to-person variation, individuals are often given both the control and the experimental treatment. The primary or null hypothesis is that the mean of the within-person differences is zero. The one-sided alternative is that the experimental treatment results are greater.

Exercise 5.18: To compare teaching methods, 10 school children were first taught by conventional methods, tested, and then taught comparable material by an entirely new approach. The following are the test results:

```
conventional 65 79 90 75 61 85 98 80 97 75
new 90 98 73 79 84 81 98 90 83 88
```

Is the new teaching method preferable to the old?

Hint: You'll need to use the R function t.test (new, conventional, "g," paired = TRUE)

5.4 COMPARING TWO SAMPLES

In this section, we'll examine the use of the binomial, Student's *t*, permutation methods, and the bootstrap for comparing two samples and then address the question of which is the best test to use. We'll also confront a problem characteristic of the applied sciences, that of developing a relevant basis for comparison.

5.4.1 What Should We Measure?

Suppose you've got this strange notion that your college's hockey team is better than mine. We compare win/lost records for last season and see that while McGill won 11 of its 15 games, your team only won 8 of 14. But is this difference statistically significant?

With the outcome of each game being success or failure, and successive games being independent of one another, it looks at first glance as if we have two series of binomial trials, but as we'll see in a moment, this is highly questionable. We could derive confidence intervals for each of the two binomial parameters. If these intervals do not overlap, then the difference in win/loss records is statistically significant. But do win/loss records really tell the story?

Let's make the comparison another way by comparing total goals. McGill scored a total of 28 goals last season and your team 32. Using the approach described in Chapter 3 for comparing two Poisson processes, we would look at this set of observations as a binomial with $28 + 32 = 60$ trials, and test the hypothesis that $p \leq 1/2$, that is, that McGill is no more likely to have scored the goal than your team, against the alternative that $p > 1/2$.

This latter approach has several problems. For one, your team played fewer games than McGill. But more telling, and the principal objection to all the methods we've discussed so far, the schedules of our two teams may not be comparable.

With binomial trials, the probability of success must be the same for each trial. Clearly, this is not the case here. We need to correct for the differences among the opponents faced by the two teams. After much discussion—what else is the off-season for?—you and I decide to award points for each game using the formula $S = O + GF - GA$, where GF stands for goals for, GA for goals against, and O is the value awarded for playing a specific opponent. In coming up with this formula and with the various values for O, we relied not on our knowledge of statistics but on our hockey expertise. This reliance on domain expertise is typical of most real-world applications of statistics.

The point totals we came up with read like this

```
McGill            4, -2, 1, 3, 5, 5, 0, -1, 6, 2, 2, 3, -2, -1, 4
Your School       3, 4, 4, -3, 3, 2, 2, 2, 4, 5, 1, -2, 2, 1.
```

Exercise 5.19: Basing your decision on the point totals, use Student's t to test the hypothesis that McGill's hockey team is superior to your school's hockey team. Is this a one-sided or two-sided test?

5.4.2 Permutation Monte Carlo

Straightforward application of the permutation methods discussed in Section 5.1.1 to the hockey data is next to impossible. Imagine how many years it would take us to look at all $\binom{14+15}{15}$ possible rearrangements! What we can do today is to look at a large random sample of rearrangements—something not possible with the primitive calculators that were all that was available in the 1930s when permutation methods were first introduced. Here's an R program that computes the original sum, then performs a *Monte Carlo*, that is, a computer simulation, to obtain the sums for 400 random rearrangements of labels:

```
➢ N = 400    #number of rearrangements to be examined
➢ sumorig = sum(McGill)
➢ n = length(McGill)
➢ cnt= 0 #zero the counter
➢ #Stick both sets of observations in a single vector
➢ A = c(McGill,Other)
➢ for (i in 1:N){
➢            D = sample (A,n)
➢            if (sum(D) <= sumorig)cnt = cnt+1
➢ }
➢ cnt/N #one-sided p-value
```

Note that the **sample()** instruction used here differs from the **sample()** instruction used to obtain a bootstrap in that the sampling in this instance is done *without* replacement.

Exercise 5.20: Show that we would have got exactly the same p-value had we used the difference in means between the samples instead of the sum of the observations in the first sample as our permutation test statistic.

*Exercise 5.21: (for mathematics and statistics majors only) Show that we would have got exactly the same p-value had we used the t-statistic as our test statistic.

These two exercises demonstrate that we can reduce the number of calculations necessary to obtain a p-value by permutation means by eliminating all sums from our test statistic that remain invariant under rearrangements of the treatment labels.

Exercise 5.22: Use the Monte Carlo approach to rearrangements to test against the alternative hypothesis that McGill's hockey team is superior to your school's hockey team.

5.4.3 One- vs. Two-Sided Tests

The preceding is an example of a *one-sided test* in which we tested a hypothesis
"The expected value of Y is not larger than the expected value of X," against a one-sided alternative, "The expected value of Y is larger than the expected value of X."

We performed the test by taking independent random samples from the X and Y populations and comparing their means. If the mean of the Y sample was less than or equal to the mean of the X sample, we accepted the hypothesis. Otherwise, we examined a series of random rearrangements in order to obtain a permutation distribution of the test statistic and a p-value.

Now suppose, we'd stated our hypothesis in the form, "The expected values of Y and X are equal," and the alternative in the form, "The expected values of Y and X are not equal." Now the alternative is two sided. Regardless of which sample mean is larger, we need to consider a random series of rearrangements. As a result, we need to double the resulting count in order to obtain a two-sided p-value.

Exercise 5.23: The following samples were taken from two different populations:
    samp1: 1.05, 0.46, 0, −0.40, −1.09, 0.88
    samp2: 0.55, 0.20, 1.52, 2.67
Were the samples drawn from two populations with the same mean?

*Exercise 5.24: Modify the R code for the Monte Carlo permutation test so as to both increase its efficiency in the one-sided case and make it applicable to either one-sided or two-sided tests.

5.4.4 Bias-Corrected Nonparametric Bootstrap

Obtaining a BCa interval for a difference in means, based on samples taken from two distinct populations, is more complicated. We need make use of an R construct known as the data frame.

Suppose you've collected the following data on starting clerical salaries in Shelby County, GA, during the periods 1975–1976 and 1982–1983.

```
$525 "female", $500 "female", $550 "female", $700
"male", $576 "female", $886 "male", $458 "female",
$600 "female", $600 "male", $850 "male", $800
"female", $850 "female", $850 "female", $800 "female"
```

and would like a confidence interval for the difference in starting salaries of men and women.

First, we enter the data and store in a data frame as follows:

```
➤ salary = c(525,500,550,700,576,886,458,600,600,850,
  800,850,850,800)
➤ sex = c(rep("female", 3), "male", "female", "male",
  "female", "female", "male", "male", c(rep("female",
  4))
➤ dframe = data.frame(sex,salary)
```

Then, we construct a function with two arguments to calculate the difference in means between bootstrap samples of men and women:

```
➤ fm.mndiff<- function(dframe, id){
    yvals<- dframe[[2]][id]
    mean(yvals[dframe[[1]]=="female"])-mean(yvals
    [dframe[[1]]=="male"])}
  }
```

The last step, computing the BCa interval, uses the same commands as in the previous chapter:

```
➤ library(boot)
➤ boot.ci(boot(dframe, fm.mndiff, 400), conf = 0.90)
```

Exercise 5.25: Compare the 90% confidence intervals for the variance of the population from which the following sample of billing data was taken for (1) the original primitive bootstrap, (2) the bias-corrected-and-accelerated nonparametric bootstrap, (3) the parametric bootstrap assuming the billing data are normally distributed, (4) the parametric bootstrap assuming the billing data are exponentially distributed.

Hospital Billing Data

```
4181,  2880,  5670,  11620,  8660,  6010,  11620,  8600,
12860, 21420, 5510,  12270,  6500,  16500, 4930,  10650,
16310, 15730, 4610,  86260,  65220, 3820,  34040, 91270,
51450, 16010, 6010,  15640,  49170, 62200, 62640, 5880,
2700,  4900,  55820, 9960,   28130, 34350, 4120,  61340,
24220, 31530, 3890,  49410,  2820,  58850, 4100,  3020,
5280,  3160,  64710, 25070
```

5.5   WHICH TEST SHOULD WE USE?

Four different tests were defined for our comparisons of two populations. Two of these were *parametric* tests that obtained their p-values by referring to parametric distributions, such as the binomial and Students *t*. Two were *resampling methods*—bootstrap and permutation test—that obtained their p-values by sampling repeatedly from the data at hand.

In some cases, the choice of test is predetermined. For example, when the observations come from or can be reduced to those that come from a binomial distribution. In other instances, as in Section 5.4.1, we need to look more deeply into the consequences of our choice. In particular, we need to consider the assumptions under which the test is valid, the effect of violations of these assumptions, and the Type I and Type II errors associated with each test.

5.5.1 *p*-Values and Significance Levels

In the preceding sections, we have referred several times to p-*values* and *significance levels*. We used both in helping us to decide whether to accept or reject a hypothesis and to take a course of action that might result in gains or losses.

To see the distinction between the two concepts, please run the following program:

```
> x = rnorm(10)
> x
> t.test(x)
> x = rnorm(10)
> x
> t.test(x)
```

There's no misprint in the above lines of R code; what this program does is take two distinct samples each of size 10 from a normal population with mean zero and unit variance, display the values of the variables in each instance, and provide two separate confidence intervals for the sample mean, one for each sample. The composition of the two samples varies, the value of the *t*-statistic varies, the p-*values* vary, and the boundaries of the confidence interval vary. What remains unchanged is the *significance level* of 100% − 95% = 5% that is used to make decisions.

You aren't confined to a 5% significance level. The R statement,

```
> t.test (x, conf.level = 0.90)
```

yields a confidence level of 90% or, equivalently, a significance level of 100% − 90% = 10%.

In clinical trials of drug effectiveness, one might use a significance level of 10% in pilot studies, but would probably insist on a significance level of 1% before investing large amounts of money in further development.

In summary, *p*-values vary from sample to sample, while significance levels are fixed.

Significance levels establish limits on the overall frequency of Type I errors. The significance levels and confidence bounds of parametric and permutation tests are *exact* only if all the assumptions that underlie these tests are satisfied. Even when the assumptions that underlie the bootstrap are satisfied, the claimed significance levels and confidence bounds of the bootstrap are only approximations. The greater the number of observations in the original sample, the better this approximation will be.

5.5.2 Test Assumptions

Virtually all statistical procedures rely on the assumption that our observations are independent of one another. When this assumption fails, the computed *p*-values may be far from accurate, and a specific significance level cannot be guaranteed.

All statistical procedures require that at least one of the following successively stronger assumptions be satisfied under the hypothesis of no differences among the populations from which the samples are drawn:

1. The observations all come from distributions that have the same value of the parameter of interest.

2. The observations are *exchangeable*, that is, each rearrangement of labels is equally likely.

3. The observations are *identically distributed* and come from a distribution of known form.

The first assumption is the weakest. If this assumption is true, a nonparametric bootstrap test* will provide an exact significance level with very large samples. The observations may come from different distributions, providing that they all have the same parameter of interest. In particular, the nonparametric bootstrap can be used to test whether the expected results are the same for two groups even if the observations in one of the groups are more variable than they are in the other.[†]

If the second assumption is true, the first assumption is also true. If the second assumption is true, a permutation test will provide exact significance levels even for very small samples.

The third assumption is the strongest assumption. If it is true, the first two assumptions are also true. This assumption must be true for a parametric test to provide an exact significance level. Indeed, to apply a parametric test when the observations come from a multiparameter distribution, such as the normal, then all population parameters, not just the one under test, must be the same for all observations under the null hypothesis. For example, a *t*-test comparing the *means* of two populations requires that the *variances* of the two populations be the same.

5.5.3 Robustness

When a test provides almost exact significance levels despite a violation of the underlying assumptions, we say that it is *robust*. Clearly, the nonparametric bootstrap is more robust than the parametric since it has fewer assumptions. Still, the parametric bootstrap, which makes more effective use of the data, will be preferable providing enough is known about the shape of the distribution from which the observations are taken. Warning: The bootstrap is not recommended for use with small samples, and for tests of parameters, which depend upon the extreme values of a distribution, the variance, for example, for which samples of 100 or more observations are required.

When the variances of the populations from which the observations are drawn are not the same, the significance level of the bootstrap used for comparing means

---

* Any bootstrap but the parametric bootstrap.
[†] We need to modify our formula and our existing R program if we suspect this to be the case; see Chapter 8.

is not affected. Bootstrap samples are drawn separately from each population. Small differences in the variances of two populations will leave the significance levels of permutation tests used for comparing means relatively unaffected but they will no longer be exact. Student's t should not be used when there are clear differences in the variances of the two groups.

On the other hand, Student's t is the exception to the rule when we say parametric tests should only be used when the distribution of the underlying observations is known. Student's-t tests for differences between means, and means, as we've already noted, tend to be normally distributed even when the observations they summarize are not.

Exercise 5.26: Repeat the following 1000 times:

a. Generate two independent samples of size 4 from a distribution that is a mixture of 0.7 $N(0,1)$ and 0.3 $N(2,1)$. Use the command: Data = rnorm(8,2 *rbinom(4,1,0.3),1)

b. Use a Monte Carlo of 400 rearrangements to derive a p-value for a test of the equality of the means of the two populations from which the samples are drawn against the one-sided alternative that the mean of the second population is larger.

c. Use a t-test to derive a p-value for a test of the equality of the means of the two populations from which the samples are drawn against the one-sided alternative that the mean of the second population is larger.

d. Plot the cumulative distribution functions of the resulting distributions of p-values side by side on a single chart and record their 1st, 5th, 10th, 90th, 95th, and 99th percentiles.

5.5.4 Power of a Test Procedure

Statisticians call the probability of rejecting the null hypothesis when an alternative hypothesis is true the *power* of the test. If we were testing a food additive for possible carcinogenic (cancer-producing) effects, this would be the probability of detecting a carcinogenic effect. The power of a test equals one minus the probability of making a Type II error. The greater the power, the smaller the Type II error, the better off we are.

Power depends on all of the following:

1. *The True Value of the Parameter Being Tested.* The greater the gap between our primary hypothesis and the true value, the greater the power will be. In our example of a carcinogenic substance, the power of the test would be greater if the substance was a strong rather than a weak carcinogen and if its effects were readily detectable.

   A word of caution: a statistically significant effect may not be of *practical significance*. The power of a test should be calculated for the smallest possible effect of practical interest.

2. **The Significance Level.** The higher the significance level (10% rather than 5%), the larger the probability of making a Type I error we are willing to accept, and the greater the probability of detecting a real effect. In our example, we would probably insist on a significance level of 1%.

3. **The Sample Size.** The larger the sample, the greater the power will be. In our example of a carcinogenic substance, the regulatory commission (the FDA in the United States) would probably insist on a power of 80%. We would then have to increase our sample size in order to meet their specifications.

4. **The Method Used for Testing.** Obviously, we want to use the most powerful possible method.

Exercise 5.27: Repeat the following 1000 times:

a. Generate a sample of size 4 from a distribution that is a mixture of 0.7 $N(0,1)$ and 0.3 $N(2,1)$. Generate a second independent sample of size 4 from a $N(2,1)$.

b. Use a Monte Carlo of 400 rearrangements to derive a p-value for a test of the equality of the means of the two populations from which the samples are drawn against the one-sided alternative that the mean of the second population is larger.

c. Use a t-test to derive a p-value for a test of the equality of the means of the two populations from which the samples are drawn against the one-sided alternative that the mean of the second population is larger.

d. Plot the cumulative distribution functions of the resulting distributions of p-values side by side on a single chart and record their 1st, 5th, 10th, 90th, 95th, and 99th percentiles.

Exercise 5.28: To test the hypothesis that consumers can't tell your cola from Coke, you administer both drinks in a blind tasting to 10 people selected at random. (1) To ensure that the probability of a Type I error is just slightly more than 5%, how many people should correctly identify the glass of Coke before you reject this hypothesis? (2) What is the power of this test if the probability of an individual correctly identifying Coke is 75%? Hint: What distribution should we make use of in determining the power?

Exercise 5.29: What is the power of the test in the preceding exercise if the probability of an individual correctly identifying Coke is 60%?

Exercise 5.30: If you test 20 people rather than 10, what will be the power of a test at the 5% significance level if the probability of correctly identifying Coke is 75%?

Exercise 5.31: Physicians evaluate diagnostic procedures on the basis of their "sensitivity" and "selectivity."

Sensitivity is defined as the percentage of diseased individuals that are correctly diagnosed as diseased. Is sensitivity related to significance level and power? If so, describe the relationship.

Selectivity, also referred to as specificity, is defined as the percentage of those diagnosed as suffering from a given disease that actually have the disease. Can selectivity be related to the concepts of significance level and power? If so, how?

Exercise 5.32: Suppose we wish to test the hypothesis that a new vaccine will be more effective than the old in preventing infectious pneumonia. We decide to inject some 1000 patients with the old vaccine and 1000 with the new and follow them for 1 year. Can we guarantee the power of the resulting hypothesis test?

Exercise 5.33: Show that the power of a test can be compared with the power of an optical lens in at least one respect.

5.6 SUMMARY AND REVIEW

In this chapter, we derived permutation, parametric, and bootstrap tests of hypothesis for comparing two samples, and for bivariate correlation. We explored the relationships and distinctions among *p*-values, significance levels, alternative hypotheses, and sample sizes. You learned that hypothesis tests and confidence intervals are intimately related. You were provided some initial guidelines to use in the selection of the appropriate test.

You made use of the R functions quantile(), cor.test(), data.frame(), and t.test().

You found that because of the variation from sample to sample, we run the risk of making one of two types of error when testing a hypothesis, each with quite different consequences. Normally, when testing hypotheses, we set a bound called the significance level on the probability of making a Type I error and devise our tests accordingly.

Exercise 5.34: Make a list of all the italicized terms in this chapter. Provide a definition for each one along with an example.

Exercise 5.35: Some authorities have suggested that when we estimate a *p*-value via a Monte Carlo as in Section 5.4.2 that we should include the original observations as one of the rearrangements. Instead of reporting the *p*-value as cnt/N, we would report it as (cnt + 1)/(N + 1). Explain why this would give a false impression. (*Hint*: Reread Chapter 2 if necessary.)

Exercise 5.36: B. Efron and R. Tibshirani (*An Introduction to the Bootstrap*, New York: Chapman & Hall, 1993) report the survival times in days for a sample of 16 mice undergoing a surgical procedure. The mice were randomly divided into two groups. The following survival times in days were recorded for a group of seven mice that received a treatment expected to prolong their survival:

➤ g.trt = c(94,197,16,38,99,141,23)

The second group of nine underwent surgery without the treatment and had these survival times in days:

➤ g.ctr = c(52,104,146,10,51,30,40,27,46)

a. Use permutation methods to test the hypothesis that the treatment does not increase survival time.

b. Provide a 75% confidence interval for the difference in mean survival days for the sampled populations based on 1000 bootstrap samples.

Exercise 5.37: Which test would you use for a comparison of the following treated and control samples?

```
control = c(4,6,3,4,7,6)
treated = c(14,6,3,12,7,15).
```

Chapter 6

Designing an Experiment or Survey

Suppose you were a consulting statistician* and were given a data set to analyze. What is the first question you would ask? "What statistic should I use?" No, your first question always should be "How were these data collected?"

Experience teaches us that garbage in, garbage out or GIGO. In order to apply statistical methods, you need to be sure that samples have been drawn at random from the population(s) you want represented and are representative of those populations. You need to be sure that the observations are independent of one another and that outcomes have not been influenced by the actions of the investigator or survey taker.

Many times people who consult statisticians don't know the details of the data collection process or they do know and look guilty and embarrassed when asked. All too often, you'll find yourself throwing your hands in the air and saying, "if only you'd come to me to design your experiment in the first place."

The purpose of this chapter is to take you step by step through the design of an experiment and a survey. You'll learn the many ways in which an experiment can go wrong. And you'll learn the right things to do to ensure your own efforts are successful.

6.1 THE HAWTHORNE EFFECT

The original objective of the industrial engineers at the Hawthorne plant of Western Electric was to see whether a few relatively inexpensive improvements would increase workers' productivity. They painted the walls green and productivity went

---

* The idea of having a career as a consulting statistician may strike you as laughable or even distasteful. I once had a student who said he'd rather eat worms and die. Suppose then that you've eaten worms and died only to wake to discover that reincarnation is real and that you've been reborn as a consulting statistician in order to expiate your sins in the previous life. I'm sure that's what must have happened in my case.

Introduction to Statistics Through Resampling Methods and R, Second Edition. Phillip I. Good.
© 2013 John Wiley & Sons, Inc. Published 2013 by John Wiley & Sons, Inc.

up. They hung posters and productivity went up. Then, just to prove how important bright paint and posters were to productivity, they removed the posters and repainted the walls a dull gray only to find that once again, *productivity went up!*

Simply put, these industrial engineers had discovered that the mere act of paying attention to a person modifies their behavior. (Note the same is true for animals.)

You've probably noticed that you respond similarly to attention from others, though not always positively. Taking a test under the watchful eye of an instructor is quite different from working out a problem set in the privacy of your room.

Physicians and witch doctors soon learn that merely giving a person a pill (any pill) or dancing a dance often results in a cure. This is called the *placebo* effect. If patients think they are going to get better, they do get better. Thus, regulatory agencies insist that before they approve a new drug, it be tested side by side with a similar-looking, similar-tasting *placebo*. If the new drug is to be taken twice a day in tablet form, then the placebo must also be given twice a day, also as a tablet, and not as a liquid or an injection. And, most important, the experimental subject should not be aware of which treatment she is receiving. Studies in which the treatment is concealed from the subject are known as *single-blind* studies.

The doctor's attitude is as important as the treatment. If part of the dance is omitted—a failure to shake a rattle, why bother if the patient is going to die anyway—the patient may react differently. Thus, the agencies responsible for regulating drugs and medical devices (in the United States, this would be the Food an Drug Administration [FDA]) now also insist that experiments be *double blind*. Neither the patient nor the doctor (or whoever administers the pill to the patient) should know whether the pill that is given the patient is an active drug or a placebo. If the patient searches the doctor's face for clues—will this experimental pill really help me?—she'll get the same response whether she is in the treatment group or is one of the *controls*.

Note: The double blind principle also applies to experimental animals. Dogs and primates are particularly sensitive to their handlers' attitudes.

6.1.1 Crafting an Experiment

In the very first set of clinical data that was brought to me for statistical analysis, a young surgeon described the problems he was having with his chief of surgery. "I've developed a new method for giving arteriograms which I feel can cut down on the necessity for repeated amputations. But my chief will only let me try out the technique on patients that he feels are hopeless. Will this affect my results?" It would and it did. Patients examined by the new method had a very poor recovery rate. But, of course, the only patients who'd been examined by the new method were those with a poor prognosis. The young surgeon realized that he would not be able to test his theory until he was able to assign patients to treatment at random.

Not incidentally, it took us three more tries until we got this particular experiment right. In our next attempt, the chief of surgery—Mark Craig of St Eligius in Boston—announced that he would do the "random" assignments. He finally was persuaded to let me make the assignment using a table of random numbers. But then he announced that he, and not the younger surgeon, would perform the operations on the patients examined by the traditional method to make sure "they were done right." Of course, this turned a comparison of methods into a comparison of surgeons and intent.

In the end, we were able to create the ideal "double-blind" study: The young surgeon performed all the operations, but the incision points were determined by his chief after examining one or the other of the two types of arteriogram.

Exercise 6.1: Each of the following studies is fatally flawed. Can you tell what the problem is in each instance and, as important, why it is a problem?

1. *Class Action.* Larry the Lawyer could barely pay his rent when he got the bright idea of looking through the county-by-county leukemia rates for our state. He called me a week later and asked what I thought of the leukemia rate in Kay County. I gave a low whistle. "Awfully high," I said.

    The next time I talked to Larry, he seemed happy and prosperous. He explained that he'd gone to Kay County once he'd learned that the principal employer in that area was a multinational chemical company. He'd visited all the families whose kids had come down with leukemia and signed them up for a class action suit. The company had quickly settled out of court when they looked at the figures.

    "How'd you find out about Kay County?" I asked.

    "Easy, I just ordered all the counties in the State by their leukemia rates and Kay came out on top."

2. *Controls.* Donald routinely tested new drugs for toxicity by injecting them in mice. In each case, he'd take five animals from a cage and inject them with the drug. To be on the safe side, he'd take the next five animals from the cage, inject them with a saline solution and use them for comparison purposes.

3. *Survey.* Reasoning, correctly, that he'd find more students home at dinnertime, Tom brought a set of survey forms back to his fraternity house and interviewed his frat brothers one by one at the dinner table.

4. *Treatment Allocation.* Fully aware of the influence that a physician's attitude could have on a patient's recovery, Betty, a biostatistician, provided the investigators in a recent clinical trial with bottles of tablets that were labeled only A or B.

5. *Clinical Trials.* Before a new drug can be marketed, it must go through a succession of clinical trials. The first set of trials (phase I) is used to establish the maximum tolerated dose. They are usually limited to 25 or so test subjects who will be observed for periods of several hours to several weeks. The second set of trials (phase II) is used to establish the minimum effective dose; they also are limited in duration and in the number of subjects involved. Only in Phase III are the trials expanded to several hundred test subjects who

will be followed over a period of months or even years. Up until the 1990s, only males were used as test subjects in order to spare women the possibility of unnecessary suffering.

6. *Comparison.* Contrary to what one would expect from the advances in medical care, there were 2.1 million deaths from all causes in the U.S. in 1985, compared to 1.7 million in 1960.

7. *Survey.* The Federal Trade Commission (FTC) surveyed former correspondence school students to see how they felt about the courses they had taken some 2–5 years earlier.* The survey was accompanied by a form letter signed by an FTC attorney that began, "The Bureau of Consumer Protection is gathering information from those who enrolled in . . . to determine if any action is warranted." Questions were multiple choice and did not include options for "I don't know" or "I don't recall."

6.2   DESIGNING AN EXPERIMENT OR SURVEY

Before you complete a single data collection form:

1. Set forth your objectives and the use you plan to make of your research.
2. Define the population(s) to which you will apply the results of your analysis.
3. List all possible sources of variation.
4. Decide how you will cope with each source. Describe what you will measure and how you will measure it. Define the experimental unit and all endpoints.
5. Formulate your hypothesis and all of the associated alternatives. Define your endpoints. List possible experimental findings along with the conclusions you would draw and the actions you would take for each of the possible results.
6. Describe in detail how you intend to draw a representative random sample from the population.
7. Describe how you will ensure the independence of your observations. Think about what might go wrong, think about what might reduce the chances of that problem happening and what you might do about it if it does.

6.2.1   Objectives

In my experience as a statistician, the people who come to consult me before they do an experiment (an all-too-small minority of my clients) aren't always clear about their objectives. I advise them to start with their reports, to write down what they would most like to see in print. For example,

---

* Macmillan, Inc. 96 F.T.C. 208 (1980).

Fifteen thousand of 17,500 surveys were completed and returned. Over half of the respondents were between the ages of 47 and 56. Thirty-six percent (36%) indicated that they were currently eligible or would be eligible for retirement in the next three years. However, only 25% indicated they intended to retire in that time. Texas can anticipate some 5,000 retirees in the next three years.

Or

743 patients self-administered our psyllium preparation twice a day over a three-month period. Changes in the Klozner-Murphy self-satisfaction scale over the course of treatment were compared with those of 722 patients who self-administered an equally foul-tasting but harmless preparation over the same time period.

All patients in the study reported an increase in self-satisfaction, but the scores of those taking our preparation increased an average of 2.3 ± 0.5 points more than those in the control group."

Adverse effects included. . . .

If taken as directed by a physician, we can expect those diagnosed with. . . .

I have my clients write in exact numerical values for the anticipated outcomes—their best guesses, as these will be needed when determining sample size. My clients go over their reports several times to ensure they've included all endpoints and as many potential discoveries as they can—"Only 25% indicated an intent to retire in that time." Once the report is fleshed out completely, they know what data needs to be collected and do not waste their time and their company's time on unnecessary or redundant effort.

Exercise 6.2: Throughout this chapter, you'll work on the design of a hypothetical experiment or survey. If you are already well along in your studies, it could be an actual one! Start now by writing the results section.

6.2.2 Sample from the Right Population

Be sure you will be sampling from the population of interest as a whole rather than from an unrepresentative subset of that population. The most famous blunder along these lines was basing the forecast of Dewey over Truman in the 1948 U.S. presidential election on a telephone survey: Those who owned a telephone and responded to the survey favored Dewey; those who voted did not.

An economic study may be flawed because we have overlooked the homeless. This was among the principal arguments the cities of New York and Los Angeles advanced against the use of the 1990 and 2000 census to determine the basis for awarding monies to cities. See City of New York v. Dept of Commerce.*

An astrophysical study was flawed because of overlooking galaxies whose central surface brightness was very low. And the FDA's former policy of permitting clinical trials to be limited to men (see Exercise 1.5) was just plain foolish.

---

* 822 F. Supp. 906 (E.D.N.Y., 1993).

Plaguing many surveys are the uncooperative and the nonresponder. Invariably, follow-up surveys of these groups show substantial differences from those who responded readily the first time around. These follow-up surveys aren't inexpensive—compare the cost of mailing out a survey to telephoning or making face-to-face contact with a nonresponder. But if one doesn't make these calls, one may get a completely unrealistic picture of how the population as a whole would respond.

Exercise 6.3: You be the judge. In each of the following cases, how would you rule?

a. The trial of *People v. Sirhan** followed the assassination of Presidential candidate Robert Kennedy. The defense appealed the guilty verdict alleging the jury was a nonrepresentative sample, offering anecdotal evidence based on the population of the Northern United States. The prosecution said, so what, our jury was representative of Los Angeles where the trial was held. How would you rule? Note that the Sixth Amendment to the Constitution of The United States provides that

A criminal defendant is entitled to a jury drawn from a jury panel which includes jurors residing in the geographic area where the alleged crime occurred.

b. In *People v. Harris*,† a survey of trial court jury panels provided by the defense showed a significant disparity from census figures. The prosecution contended that the survey was too limited, being restricted to the Superior Courts in a single district, rather than being county wide. How would you rule?

c. Amstar Corporation claimed that "Domino's Pizza" was too easily confused with its own use of the trademark "Domino" for sugar.‡ Amstar conducted and offered in evidence a survey of heads of households in ten cities. Domino objected to this survey pointing out that it had no stores or restaurants in eight of these cities and, in the remaining two their outlets had been open less than three months. Domino provided a survey it had conducted in its pizza parlors and Amstar objected. How would you rule?

Exercise 6.4: Describe the population from which you plan to draw a sample in your hypothetical experiment. Is this the same population you would extend the conclusions to in your report?

---

* 7 Cal.3d 710, 102 Cal. Rptr.385 (1972), *cert. denied*, 410 U.S. 947.
† 36 Cal.3d 36, 201 Cal. Rptr 782 (1984), *cert. denied* 469 U.S. 965, *appeal to remand* 236 Cal. Rptr 680, 191 Cal. App. 3d 819, *appeal after remand*, 236 Cal. Rptr 563, 217 Cal. App. 3d 1332.
‡ Amstar Corp. v. Domino's Pizza, Inc., 205 U.S.P.Q 128 (N.D. Ga. 1979), *rev'd*, 615 F. 2d 252 (5th Cir. 1980).

> ### The Drunk and the Lamppost
>
> There's an old joke dating back to at least the turn of the previous century about the drunk whom the police officer found searching for his wallet under the lamppost. The policeman offers to help and after searching on hands and knees for 15 minutes without success asks the inebriated gentleman just exactly where he lost his wallet. The drunk points to the opposite end of the block. "Then why were you searching over here?!" the policeman asks.
>
> "The light's better."
>
> It's amazing how often measurements are made because they are convenient (inexpensive and/or quick to make) rather than because they are directly related to the object of the investigation. Your decisions as to what to measure and how to measure it require as much or more thought as any other aspect of your investigation.

6.2.3 Coping with Variation

As noted in the very first chapter of this text, you should begin any investigation where variation may play a role by listing all possible sources of variation—in the environment, in the observer, in the observed, and in the measuring device. Consequently, you need to have a thorough understanding of the domain—biological, psychological, or seismological in which the inquiry is set.

Will something as simple as the time of day affect results? Body temperature and the incidence of mitosis both depend on the time of day. Retail sales and the volume of mail both depend on the day of the week. In studies of primates (including you) and hunters (tigers, mountain lions, domestic cats, dogs, wolves, etc.), the sex of the observer will make a difference.

Statisticians have found four ways for coping with individual-to-individual and observer-to-observer variation:

1. *Controlling.* Making the environment for the study—the subjects, the manner in which the treatment is administered, the manner in which the observations are obtained, the apparatus used to make the measurements, and the criteria for interpretation—as uniform and homogeneous as possible.

2. *Blocking.* A clinician might stratify the population into subgroups based on such factors as age, sex, race, and the severity of the condition and restrict subsequent comparisons to individuals who belong to the same subgroup. An agronomist would want to stratify on the basis of soil composition and environment.

3. *Measuring.* Some variables, such as cholesterol level or the percentage of CO_2 in the atmosphere, can take any of a broad range of values and don't lend themselves to blocking. As we show in the next chapter, statisticians have methods for correcting for the values taken by these *covariates*.

4. *Randomizing.* Randomly assign patients to treatment within each block or subgroup so that the innumerable factors that can neither be controlled nor observed directly are as likely to influence the outcome of one treatment as another.

Exercise 6.5: List all possible sources of variation for your hypothetical experiment and describe how you will cope with each one.

6.2.4 Matched Pairs

One of the best ways to eliminate a source of variation and the errors in interpretation associated with it is through the use of matched pairs. Each subject in one group is matched as closely as possible by a subject in the treatment group. If a 45-year-old black male hypertensive is given a blood-pressure lowering pill, then we give a second similarly built 45-year-old black male hypertensive a placebo.

Consider the case of a fast-food chain that is interested in assessing the effect of the introduction of a new sandwich on over-all sales. To do this experiment, they designate a set of outlets in different areas—two in the inner city, two in the suburbs, two in small towns, and two located along major highways. A further matching criterion is that the overall sales for the members of each pair prior to the start of the experiment were approximately the same for the months of January through March. During the month of April, the new sandwich is put on sale at one of each pair of outlets. At the end of the month, the results are recorded for each matched pair of outlets (Table 6.1).

To analyze these data, we consider the 28 possible rearrangements that result from the possible exchanges of labels within each matched pair of observations.

```
➤ New = c(48722, 28965, 36581, 40543, 55423, 38555,
  31778, 45643)
➤ Standard = c(46555, 28293, 37453, 38324, 54989,
  35687, 32000, 43289)
➤ N = 400 #number of rearrangements to be examined
➤ sumorig = sum(New)
➤ n=length(New)
➤ stat = numeric (n)
➤ cnt = 0 #zero the counter
➤ for (i in 1:N){
+     which = runif(n)
+     for (j in 1:n) stat[j] = ifelse
        (runif(1) < 0.5, New[j], Standard[j])
+     if (sum(stat) <= sumorig)cnt = cnt+1
➤ }
➤ cnt/N #one-sided p-value
[1] 0.98
```

Exercise 6.6: In the preceding example, was the correct *p*-value 0.98, 0.02, or 0.04? Hint: What was the principle alternative? Was it one-sided or two-sided?

Table 6.1   Sales (in $) of New Sandwich during April

	1	2	3	4	5	6	7	8
New menu	48722	28965	36581	40543	55423	38555	31778	45643
Standard	46555	28293	37453	38324	54989	35687	32000	43289

Exercise 6.7: Did the increased sales for the new menu justify the increased cost of $1200 per location?

6.2.5   The Experimental Unit

A scientist repeatedly subjected a mouse named Harold to severe stress. She made a series of physiological measurements on Harold, recording blood pressure, cholesterol levels, and white blood cell counts both before and after stress was applied for a total of 24 observations. What was the sample size?

Another experimenter administered a known mutagen—a substance that induces mutations, into the diet of a pregnant rat. When the rat gave birth, the experimenter took a series of tissue samples from each of the seven offspring, two from each of eight body regions. What was the sample size?

In each of the preceding examples, the sample size was one. In the first example, the sole *experimental unit* was Harold. In the second example, the experimental unit was a single pregnant rat. Would stress have affected a second mouse the same way? We don't know. Would the mutagen have caused similar damage to the offspring of a different rat? We don't know. We do know there is wide variation from individual to individual in their responses to changes in the environment. With data from only a single individual in hand, I'd be reluctant to draw any conclusions about the population as a whole.

Less obvious is that when the observations are made on a subgroup with common interests such as Marine volunteers, physics graduate students, all the residents of a gated community, or new mothers, the responses within the subgroup may be highly correlated. Depending on the object of the investigation, it may be necessary to treat all the observations on such a cluster as if they were a single observation. (See Section 6.2.11.)

Exercise 6.8: Suppose we are testing the effect of a topical ointment on pink eye. Is each eye a separate experimental unit or is each patient?

6.2.6   Formulate Your Hypotheses

In translating your study's objectives into hypotheses that are testable by statistical means, you need to satisfy all of the following:

- The hypothesis must be numeric in form and must concern the value of some population parameter. Examples: More than 50% of those registered to vote in the State of California prefer my candidate. The arithmetic average of errors in tax owed that are made by U.S. taxpayers reporting $30,000–$50,000 dollars income is less than $50. The addition of vitamin E to standard cell growth medium will increase the lifespan of human diploid fibroblasts by no less than 30 generations. Note in these examples that we've also tried to specify the population from which sample are to be taken as precisely as possible.

- There must be at least one meaningful numeric alternative to your hypothesis.

- It must be possible to gather data to test your hypothesis.

The statement, "redheads are sexy" is not a testable hypothesis. Nor is the statement, "everyone thinks redheads are sexy." Can you explain why? The statement "at least 80% of Reed College students think redheads are sexy," is a testable hypothesis.

You should also decide at the same time as you formulate your hypotheses whether the alternatives of interest are one-sided or two-sided, ordered or unordered.

Exercise 6.9: Are the following testable hypotheses? Why or why not?

a. A large meteor hitting the Earth would dramatically increase the percentage of hydrocarbons in the atmosphere.
b. Our candidate can be expected to receive votes in the coming election.
c. Intelligence depends more on one's genes than on one's environment.

6.2.7 What Are You Going to Measure?

In order to formulate a hypothesis that is testable by statistical means, you need decide on the variables you plan to measure. Perhaps, your original hypothesis was that men are more intelligent than women. To put this in numerical terms requires a scale by which intelligence may be measured. Which of the many scales do you plan to use, and is it really relevant to the form of intelligence you had in mind?

Be direct. To find out which drugs individuals use and in what combinations, which method would yield more accurate data: (1) a mail survey of households, (2) surveying customers as they step away from a pharmacy counter, or (3) accessing pharmacy records?

Clinical trials often make use of *surrogate response variables* that are less costly or less time-consuming to measure than the actual variable of interest. One of the earliest examples of the use of a surrogate variable was when coal miners would take a canary with them into the mine to detect a lack of oxygen well before the miners themselves fell unconscious. Today, with improved technology, they would be able to measure the concentration of oxygen directly.

The presence of HIV virus often serves as a surrogate for the presence of AIDS. But is HIV an appropriate surrogate? Many individuals have tested positive for HIV who do not go on to develop AIDS.* How shall we measure the progress of arteriosclerosis? By cholesterol levels? Angiography? Electrocardiogram? Or by cardiovascular mortality?

> **Exercise 6.10:** Formulate your hypothesis and all of the associated alternatives for your hypothetical experiment. Decide on the variables you will measure. List possible experimental findings along with the conclusions you would draw and the actions you would take for each possible outcome. (A spreadsheet is helpful for this last.)

6.2.8 Random Representative Samples

Is randomization really necessary? Would it matter if you simply used the first few animals you grabbed out of the cage as controls? Or if you did all your control experiments in the morning and your innovative procedures in the afternoon? Or let one of your assistants perform the standard procedure while you performed and perfected the new technique?

A sample consisting of the first few animals to be removed from a cage will not be random because, depending on how we grab, we are more likely to select more active or more passive animals. Activity tends to be associated with higher levels of corticosteroids, and corticosteroids are associated with virtually every body function.

We've already discussed in Section 6.1.1 why a simple experiment can go astray when we *confound* competing sources of variation, such as the time of day, and the observer with the phenomena that is our primary interest. As we saw in the preceding section, we can block our experiment and do the control and the innovative procedure both in the afternoon and in the morning, but we should not do one at one time and one at the other. Recommended in the present example would be to establish four different blocks (you observe in the morning, you observe in the afternoon, your assistant observes in the morning, your assistant observes in the afternoon) and to replicate the experiment separately in each block.

Samples also are taken whenever records are audited. Periodically, federal and state governments review the monetary claims made by physicians and Health Maintenance Organizations for accuracy. Examining each and every claim would be prohibitively expensive, so governments limit their audits to a sample of claims. Any systematic method of sampling, examining every 10th claim say, would fail to achieve the desired objective. The HMO would soon learn to maintain its files in an equally systematic manner, making sure that every 10th record was error-free and fraud-free. The only way to ensure honesty by all parties submitting claims is to let a sequence of random numbers determine which claims will be examined.

---

* A characteristic of most surrogates is that they are not one-to-one with the gold standard.

The same reasoning applies when we perform a survey. Let us suppose we've decided to subdivide (block) the population whose properties we are investigating into strata—males, females, city dwellers, farmers—and to draw separate samples from each stratum. Ideally, we would assign a random number to each member of the stratum and let a computer's random number generator determine which members are to be included in the sample.

By the way, if we don't *block* our population, we run the risk of obtaining a sample in which members of an important subgroup are absent or underrepresented. Recall from Section 2.2.3, that a single jury (or sample) may not be representative of the population as a whole. We can forestall this happening by deliberately drawing samples from each important subgroup.*

Exercise 6.11: Suppose you were to conduct a long-term health survey of our armed services personnel. What subgroups would you want to consider? Why?

Exercise 6.12: Show that the size of each subsample need not be proportional to its size in the population at large. For example, suppose your objective was to estimate the median annual household income for a specific geographic area and you were to take separate samples of households whose heads were male and female, respectively. Would it make sense to take samples of the same size from each group?

6.2.9 Treatment Allocation

If the members of a sample taken from a stratum are to be exposed to differing test conditions or treatments, then we must make sure that treatment allocation is random and that the allocation is concealed from both the investigator and the experimental subjects in so far as this is possible.

Treatment allocation cannot be left up to the investigator because of the obvious bias that would result. Having a third party label the treatments (or the treated patients) with seemingly meaningless labels, such as A or B, won't work either. The investigator will soon start drawing conclusions—not necessarily the correct ones—about which treatment the A's received. In clinical trials, sooner or later, the code will need to be broken for a patient who is exhibiting severe symptoms that require immediate treatment. Breaking the code for one patient when the A/B method of treatment allocation is used will mean the code has been broken for all patients.

Similar objections can be made to any system of treatment allocation in which subjects are assigned on a systematic basis to one treatment regimen or the other. For example, injecting the first subject with the experimental vaccine, the next with saline, and so forth. The only safe system is one in which the assignment is made on the basis of random numbers.

* There has been much argument among legal scholars as to whether such an approach would be an appropriate or constitutional way to select juries.

6.2.10   Choosing a Random Sample

Exercise 6.13: Describe the method of sampling you will use in your hypothetical experiment.

   If you already have a set of population data in hand, we'll show you in the next chapter how to use R select a random subset from those data.

Exercise 6.14: Once again, you be the judge. The California Trial Jury Selection and Management Act* states that

It is the policy of the State of California that all persons selected for jury service shall be selected at random from the population of the area served by the court; that all qualified persons have an equal opportunity, in accordance with this chapter, to be considered for jury service in the state and an obligation to serve as jurors when summoned for that purpose; and that it is the responsibility of jury commissioners to manage all jury systems in an efficient, equitable, and cost-effective manner in accordance with this chapter.

In each of the following cases, decide whether this Act has been complied with.

1. A trial judge routinely excuses physicians from jury duty because of their importance to the community.

2. Jury panels are selected from lists of drivers compiled by the Department of Motor Vehicles.†

3. A trial judge routinely excuses jurors not possessing sufficient knowledge of English.‡

4. A trial judge routinely excuses the "less educated," (12 or less years of formal education) or "blue-collar workers."§

5. A trial judge routinely excuses anyone who requests to be excused.

6. Jury selection is usually a two- or three-stage process. At the first stage, a panel is selected at random from the population. At the second stage, jurors are selected from the panel and assigned to a courtroom. In People v. Viscotti,** the issue was whether the trial court erred in taking the first 12 jurors from the panel rather than selecting 12 at random. How would you rule?

7. A jury of 12 black males was empanelled in an area where blacks and whites were present in equal numbers.

6.2.11   Ensuring Your Observations Are Independent

Independence of the observations is essential to most statistical procedures. Any kind of dependence, even if only partial, can make the analysis suspect.

---

* Title 3, C.C.P. Section 191.
† U.S. v. Bailey, 862 F. Supp. 277 (D. Colo. 1994) *aff'd in part, rev'd in part*, 76 F.3d 320, *cert. denied* 116 S. Ct. 1889.
‡ People v. Lesara, 206 Cal. App. 3d 1305, 254 Cal. Rptr. 417 (1988).
§ People v. Estrada, 93 Cal. App. 3d 76, 155 Cal. Rptr. 731 (1979).
** 2 Cal. 4th 1, 5 Cal. Rptr.2d 495 (1992).

Too often, surveys take advantage of the cost savings that result from naturally occurring groups such as work sites, schools, clinics, neighborhoods, even entire towns or states. Not surprisingly, the observations within such a group are correlated. Any group or cluster of individuals who live, work, study, or pray together may fail to be representative for any or all of the following reasons:

- Shared exposure to the same physical or social environment
- Self-selection in belonging to the group
- Sharing of behaviors, ideas, or diseases among members of the group.

Two events A and B are independent only if knowledge of B is NEVER of value in predicting A. In statistics, two events or two observations are said to be independent if knowledge of the outcome of the one tells you nothing about the likely outcome of the other. My knowledge of who won the first race will not help me predict the winner of the second. On the other hand, knowledge of the past performance of the horses in the second race would be quite helpful.

The UCLA statistics professor who serves as consultant to the California State lottery assures me that the winning numbers on successive days are completely independent of one another. Despite the obvious, the second most common pick in the California lottery are the numbers that won the previous day!

Pick two voters at random and the knowledge of one person's vote won't help me forecast the others. But if you tell me that the second person is the spouse of the first, then there is at least a partial dependence. (The two spouses may vote differently on specific occasions, but if one generalizes to all spouses on all occasions, the dependence is obvious.) The effect of such correlation must be accounted for by the use of the appropriate statistical procedures.*

The price of Coca Cola stock tomorrow does depend upon the closing price today. But the change in price between today and tomorrow's closing may well be independent of the change in price between yesterday's closing and today's. When monitoring an assembly line or a measuring instrument, it is the changes from hour to hour and day to day that concern us. Change is expected and normal. It is the trends in these changes that concern us as these may indicate an underlying mutual dependence on some other hidden factors.

Exercise 6.15: Review Exercise 2.20.

6.3   HOW LARGE A SAMPLE?

Once we have mastered the technical details associated with making our observations, we are ready to launch our experiment or survey, but for one unanswered question: How large a sample should we take?

---

* e.g., Z. Feng et al., Selected statistical issues in group randomized trials, *Annual Rev. Public Health* 2001; 22: 167–187.

The effect of increasing sample size is best illustrated in the following series of photographs supplied by OZTAM. The picture (below) is comprised of several hundred thousand tiny dots (the population).

Now suppose we were to take successive representative samples from this population consisting of 250, 1000, and 2000 dots, respectively. They are "area probability" samples of the original picture, because the dots are distributed in proportion to their distribution in the picture. If we think of homes instead of dots, this is the sampling method used for most door-to-door surveys.

Having trouble recognizing the photo? Move back 30 in or so from the page. When your eye stops trying to read the dots, even the smallest sample provides a recognizable picture. You would have trouble picking this woman out of a group based on the 250-dot sample. But at 1000 dots, if you squint to read the pattern of light and dark, you might recognize her. At 2000 dots, you see her more clearly—but the real improvement is between 250 and 1000—an important point. In sampling, the ability to see greater detail is a "squared function"—it takes four times as large a sample to see twice the detail. This is the strength and weakness of sample-based research. You can get the general picture cheap, but precision costs a bundle.

In our hypothetical experiment, we have a choice of using either a sample of fixed size, or of a sequential sampling method in which we proceed in stages, deciding at each stage whether to terminate the experiment and make a decision. In the balance of this chapter, we shall focus on methods for determining a fixed sample size, merely indicating some of the possibilities associated with sequential sampling.

6.3.1   Samples of Fixed Size

Not surprisingly, many of the factors that go into determining optimal sample size are identical with those needed to determine the power of a test (Section 5.5.4):

1. The true value of the parameter being tested—the greater the gap between our primary hypothesis and the true value, the smaller the sample needed to detect the gap.

2. The variation of the observations. The more variable the observations, the more observations we will need to detect an effect of fixed size.

3. The significance level and the power. If we fix the power against a specific alternative, then working with a higher significance level (10% rather than 5%) will require fewer observations.

4. The relative costs of the observations and of the losses associated with making Type I and Type II errors. If our measurements are expensive, then to keep the over-all cost of sampling under control, we may have to accept the possibility of making Type I and Type II errors more frequently.

5. The method used for testing. Obviously, we want to use the most powerful possible method to reduce the number of observations.

The sample size that we determine by consideration of these factors is the sample size we need to end the study with. We may need to take a much larger sample to begin with in order to account for dropouts and withdrawals, animals that escape from their cages, get mislabeled or misclassified, and so forth. Retention is a particular problem in long-term studies. In a follow-up survey conducted 5 years after the original, one may be able to locate as few as 20% of the original participants.

We have a choice of methods for determining the appropriate sample size. If we know how the observations are distributed, we should always take advantage of our knowledge. If we don't know the distribution exactly, but the sample is large enough that we feel confident that the statistic we are interested in has almost a normal distribution, then we should take advantage of this fact. As a method of last resort, we can run a simulation or bootstrap.

In what follows, we consider each of these methods in turn.

6.3.1.1   Known Distribution

One instance in which the distribution of the observations is always known is when there are only two possible outcomes—success or failure, a vote for our candidate or a vote against. Suppose we desire to cut down a stand of 500-year-old redwoods in order to build and sell an expensive line of patio furniture. Unfortunately, the stand is located on State property, but we know a politician who we feel could be persuaded to help facilitate our purchase of the land. In fact, his agent has provided us with a rate card.

The politician is up for re-election and believes that a series of TV and radio advertisements purchased with our money could guarantee him a victory. He claims that those advertisements would guarantee him 55% of the vote. Our own advisors say he'd be lucky to get 40% of the vote without our ads, and the best the TV exposure could do is give him another 5%.

We decide to take a poll. If it looks like only 40% of the voters favor the candidate, we won't give him a dime. If 46% or more of the voters already favor him, we'll pay to saturate the airwaves with his promises. We decide we can risk making a Type I error 5% of the time and a Type II error at most 10% of the time. That is, if $p = 0.40$, then the probability of rejecting the hypothesis that $p = 0.40$ should be no greater than 5%. And if $p = 0.46$, then the probability of rejecting the hypothesis that $p = 0.40$ should be at least 90%.

```
➤ #Calculate the 95% percentile of the binomial
  #distribution with 10 trials and p = 0.4
➤ qbinom(.95,10,.4)
[1] 7
```

The quantile function qbinom always gives the next largest integer quantity. Let's see what Type I error would be associated with rejecting for values larger than 6.

```
#Calculate the probability of observing more than 6
#successes for a binomial distribution with 10 trials
#and p = 0.4
➤ pbinom(6,10,.4, FALSE)
[1] 0.05476188
```

Excellent, now let's see what the Type II error would be:

```
#Calculate the probability of observing 6 or less
#successes for a binomial distribution with 10 trials
#and p = 0.46
➤ pbinom(6,10,.46)
[1] 0.8859388
```

Much too large, let's try a slightly larger sample size:

```
➤ qbinom(.95,100,.4)
[1] 48
➤ pbinom(47,100,.4, FALSE)
[1] 0.06378918
```

Still a little large. The Type II error is far too large

```
➤ pbinom(47,100,.46)
[1] 0.6191224
```

Let's try a much larger sample size of 400

```
➤ qbinom(.95,400,.4)
[1] 176
```

The Type I error is

```
➤ pbinom(175,400,.4, FALSE)
[1] 0.05734915
```

The Type II error is

```
➢ pbinom(175,400,.46)
[1] 0.1970185
```

We're getting close. Let's try a sample size of 800.

```
➢ qbinom(.95,800,.4)
[1] 343
➢ pbinom(342,800,.40, FALSE)
[1] 0.0525966
➢ pbinom(342,800,.46)
[1] 0.03503062 which is less than 10%.
```

A poll of 800 people will give me the results I need. But why pay for that large a sample, when fewer observations will still result in the desired Type I and Type II errors.

Exercise 6.16: Find to the nearest 20 observations the smallest sample size needed to yield a Type I error of 5% when $p = 0.40$ and a Type II error of 10% when $p = 0.46$.

Exercise 6.17: A friend of ours has a "lucky" coin that seems to come up heads every time he flips it. We examine the coin and verify that one side is marked tails. How many times should we flip the coin to test the hypothesis that it is fair so that (1) the probability of making a Type I error is no greater than 10% and (2) we have a probability of 80% of detecting a weighted coin that will come up heads 70 times out of one hundred on the average?

6.3.1.2   Almost Normal Data

As noted in Chapter 4, the mean of a sample will often have an almost normal distribution similar to that depicted in Figure 1.4 even when the individual observations come from some quite different distribution. See, for example, Exercise 4.3.

In the previous section, we derived the ideal sample size more or less by trial and error. We could proceed in much the same way with normally distributed data, but there is a much better way. Recall that in Exercise 4.11, we showed that the variance of the mean of n observations each with variance σ^2 was σ^2/n. If we know the cut-off value for testing the mean of a normal distribution with variance 1, we can find the cutoff value and subsequently the power of a test for the mean of a normal distribution with any variance whatever.

Let's see how. Typing

```
➢ qnorm(.95)
```

we find that the 95th percentage point of an $N(0,1)$ distribution is

```
[1] 1.644854
```

We illustrate this result in Figure 6.1.

If the true mean is actually one standard deviation larger than 0, the probability of observing a value larger than 1.644854 is given by

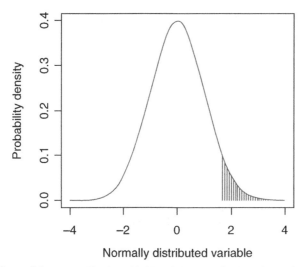

Figure 6.1   A cut-off value of 1.64 excludes 5% of N(0,1) observations.

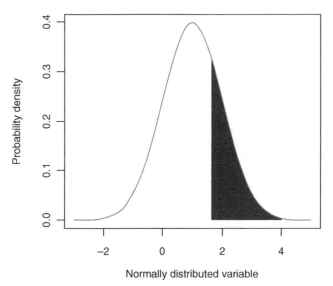

Figure 6.2   A cut-off value of 1.64 detects 26% of N(1,1) observations.

```
➤ pnorm(1.644854,1,1, FALSE)
```

or equivalently

```
1 - pnorm(1.644854,1)
[1] 0.2595109
```

We illustrate this result in Figure 6.2.

We can use Exercise 4.11 to show that if each of n independent observations is normally distributed as N(0,1), then their mean is distributed as N(0,1/n). Detecting a difference of 1 now becomes a task of detecting a difference of $\sqrt{n}$ standard deviation units. We will have power of 90% providing 1—pnorm(1.644854, sqrt(n)) = 0.1.

```
➢ qnorm(.1)
[1] -1.281552
```

Thus, we require a sample size n such that $1.644854 - \sqrt{n} = -1.281552$ or $n = 9$.

More often, we need to test against an alternative of fixed size. For example, the population mean is really equal to 10 units. Thus, to determine the required sample size for testing, we would also need to know the population variance or at least to have some estimate of what it is. If we have taken a preliminary sample, then the variance of this sample s^2 could serve as an estimate of the unknown population variance σ^2.

Let us suppose the sample variance is 25. The sample standard deviation is 5, and we are testing for an estimated difference of 2 standard deviations. At a significance level of 5%, we require a sample size n such that $1.644854 - \sqrt{n}*10/5 = -1.281552$ or no more than three observations.

To generalize the preceding results, suppose that C_α is the cut-off value for a test of significance at level α, and we want to have power β to detect a difference of size δ. C_β is the value of a $N(0,1)$ distributed variable Z for which $P\{Z > C_\beta\} = \beta$, and σ is the standard deviation of the variable we are measuring. Then $\sqrt{n} = (C_\alpha - C_\beta)\sigma/\delta$.

> **Exercise 6.18:** How big a sample would you need to test the hypothesis that the average sixth grader is 150cm in height at a 5% significance level so that the probability of detecting a true mean height of 160cm is 90%? (Take advantage of my classroom data.)

> **Exercise 6.19:** When his results were not statistically significant at the 10% level, an experimenter reported that a "new experimental treatment was ineffective." What other possibility is there?

6.3.1.3 Bootstrap

If you have reason to believe that the distribution of the sample statistic is not normal, for example, if you are testing hypotheses regarding variances or ratios, the best approach to both power and sample size determination is to bootstrap from the empirical distribution under both the primary and the alternative hypothesis.

Recently, one of us was helping a medical device company design a comparison trial of their equipment with that of several other companies. They had plenty of information on their own equipment but could only guess at the performance characteristics of their competitors. As they were going to have to buy, then destroy their competitors' equipment to perform the tests, they wanted to keep the sample size as small as possible.

They performed a series of stress tests; the resultant scores took values from 0 to 5, with 5 being the best. The idea was to take a sample of k units from each lot and reject if the mean score was too small. To determine the appropriate cut-off value for each prospective sample size k, we took a series of simulated samples from the empirical distribution for our client's equipment. The individual frequencies for this distribution were $f0$, $f1$, $f2$, $f3$, $f4$, and $f5$. We let the computer choose a random number from 0 to 1. If this number was less than $f0$, we set the simulated test score to 0. If the random number was greater than f0 but less than $f0 + f1$, we set it to 1, and so forth. We did this k times, recorded the mean, and then repeated the entire process MC times to obtain a distribution of possible outcomes. For $k = 4$, 95% of the time this mean was greater than 3. So 3 was our cutoff point for $k = 4$.

Here is the R code for the preceding calculations:

```
freq = c(0.06,0.12,0.18,0.24,0.31,0.09)
n = length(freq)
cfreq = numeric(n)
cfreq[1] = freq[1]
for(i in 2:n)cfreq[i] = cfreq[i-1]+freq[i]
k = 4
MC = 100
dist = numeric(MC)
#Draw MC random samples
for (m in 1:MC){
    scr = runif(k)
    sum = 0
    for (i in 1:k){
        scrp = scr[i]
        for (j in 1:n){
            if(scrp< = cfreq[j]){
                tmp = j-1
                break
            }
        }
    sum = sum+tmp
    }
dist[m] = sum/k
}
```

Next, we guesstimated an empirical distribution for the competitor's product. We repeated the entire simulation using this guesstimated distribution from which to draw our samples. (Sometimes, you just have to reply on your best judgment.) For $k = 4$, 40% of the time the mean of our simulated samples of the competitors' products was less than 3. Not good enough. We wanted a test our product could pass and their products wouldn't.

By trial and error, we finally came up with a sample size of 6 and a test our product could pass 95% of the time and the competitors' products would fail at least 70% of the time. We were happy.

How many simulated samples n did we have to take each time? The proportion of values that fall into the rejection region is a binomial random variable with n trials and a probability β of success in each trial, where β is the desired power of the test.

We used $n = 100$ until we were ready to fine-tune the sample size, when we switched to $n = 400$.

Exercise 6.20: In preliminary trials of a new medical device, test results of 7.0 were observed in 11 out of 12 cases and 3.3 in 1 out of 12 cases. Industry guidelines specified that any population with a mean test result greater than 5 would be acceptable. A worst -case or boundary-value scenario would include one in which the test result was 7.0, 3/7th of the time, 3.3, 3/7th of the time, and 4.1, 1/7th of the time.

The statistical procedure with significance level of 6% requires us to reject if the sample mean of subsequent test results is less than 6. What sample size is required to obtain a power of at least 80% for the worst-case scenario?

6.3.2 Sequential Sampling

The computational details of sequential sampling procedures are beyond the scope of the present text. Still, realizing that many readers will go on to design their own experiments and surveys, we devote the balance of this chapter to outlining some of the possibilities.

6.3.2.1 Stein's Two-Stage Sampling Procedure

Charles Stein's two-stage sampling procedure makes formal recognition of the need for some estimate of variation before we can decide on an optimal sample size. The procedure assumes that the test statistic will have an almost normal distribution. We begin by taking a relatively small sample and use it and the procedures of the pre-ceding sections to estimate the optimal sample size.

If the estimated optimal sample size is less than or equal to the size of the sample we've already taken we stop, otherwise we will take the number of observations suggested by the data we've just gathered plus one.

Exercise 6.21: Apply Stein's two-stage sampling procedure to the data of Exercise 6.17. How many additional observations would we need to detect an improvement in scores of 4 units 95% of the time?

6.3.2.2 Wald Sequential Sampling

When our experiments are destructive in nature (as in testing condoms) or may have an adverse effect upon the experimental subject (as in clinical trials), we would prefer not to delay our decisions until some fixed sample size has been reached.

Figure 6.3 depicts a sequential trial of a new vaccine after five patients who had received either the vaccine or an innocuous saline solution had come down with the disease. Each time a patient came down with the disease, the jagged line was extended to the right. Each time a patient who had received the saline solution came down with the disease, the jagged line was extended downward one notch. Each time a patient who had received the experimental vaccine came down with the

Sequential boundaries using the Lan–DeMets method

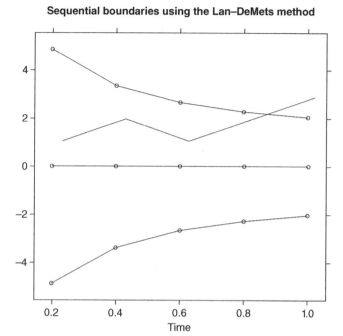

Figure 6.3 Sequential trial in progress. Scales refer to number of cases of the disease.

disease, the jagged line was extended upward one notch. The experiment will continue until either the jagged line crosses the lower boundary—in which case, we will stop the experiment, reject the null hypothesis and immediately put the vaccine into production, or the jagged line crosses the upper boundary—in which case, we will stop the experiment, accept the null hypothesis and abandon further work with this vaccine. What Abraham Wald showed in his pioneering research was that on the average, the resulting sequential experiment would require many fewer observations whether or not the vaccine was effective than would a comparable experiment of fixed sample size.

Exercise 6.22: Suppose we were to take a series of observations and, after each one, reject if the test statistic were greater than the 95th percentile of its distribution under the null hypothesis. Show that the Type I error would exceed 5% even if we only took two observations.

As Exercise 6.22 illustrates, simplify performing a standard statistical test after each new observation as if the sample size were fixed will lead to inflated values of Type I error. The boundaries depicted in Figure 6.3 were obtained using formulas specific to sequential design. Not surprisingly, these formulas require us to know each and every one of the factors required to determine the number of samples when an experiment is of fixed size.

More recent developments include "group sequential designs," which involve testing not after every observation as in a fully sequential design, but rather after groups of observations, for example, after every 6 months in a clinical trial.

To design a group sequential trial, follow the procedure described in Section 4.3.2 to download and install the **ldbound** package. Remember, each time you wish to use this package, you must type the command

> ➢ `library (ldbound)`

The principal R function in this package, bound(), which we used to construct Figure 6.3, has a large number of options; be sure to consult its instruction manual via R's Help command (see Figure 6.4) before you attempt its use.

Figure 6.5 depicts the boundaries of a group sequential trial at each of four analysis times. At each of these times, a difference in proportions that falls below the lower boundary or above the upper boundary will cause the trial to stop; anything in the middle causes it to continue.

For comparison, a fixed trial (in which one only analyzes the data at the completion of the study) is shown in the figure; such a trial would require just under 6000 subjects for the same Type I error and power. Note that in almost every instance, except when there is virtually no difference between the two treatments, the sequential method requires fewer observations.

Figure 6.4   Accessing R's help menu.

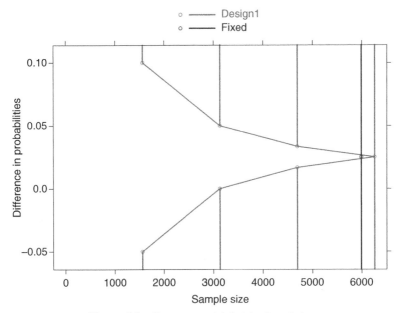

Figure 6.5 Group-sequential decision boundaries.

The major benefit of sequential designs is that we may stop early if results clearly favor one or the other hypothesis. For example, if the treatment really is worse than the control, we are likely to hit one of the lower boundaries early. If the treatment is much better than the control, we are likely to hit an upper boundary early. Even if the true difference is right in the middle between our two hypotheses, say that the treatment is 2.5% better (when the alternative hypothesis is that it is 5% better), we may stop early on occasion. Figure 6.6 shows the average sample size as a function of theta, the true difference in means. When theta is less than 0% or greater than 5%, we need about 4000 observations on average before stopping. Even when the true difference is right in the middle, we stop after about 5000 observations, on average. In contrast, the fixed-sample design requires nearly 6000 observations for the same Type I error and power.

6.3.2.3 Adaptive Sampling

The adaptive method of sequential sampling is used primarily in clinical trials where either the treatment or the condition being treated presents substantial risks to the experimental subjects. Suppose, for example, 100 patients have been treated, 50 with the old drug and 50 with the new. If on reviewing the results, it appears that the new experimental treatment offers substantial benefits over the old, we might change the proportions given each treatment, so that in the next group of 100 patients, just 25 randomly chosen patients receive the old drug and 75 receive the new.

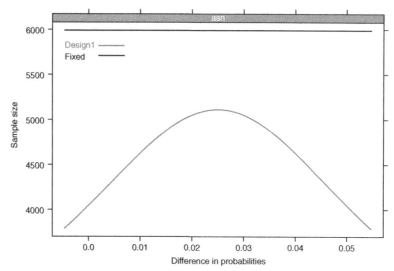

Figure 6.6   Average sample sizes, for group-sequential design.

6.4   META-ANALYSIS

Such is the uncertain nature of funding for scientific investigation, that experimenters often lack the means necessary to pursue a promising line of research. A review of the literature in your chosen field is certain to turn up several studies in which the results are inconclusive. An experiment or survey has ended with results that are "almost" significant, say with $p = 0.075$ but not $p = 0.049$. The question arises whether one could combine the results of several such studies thereby obtaining, in effect, a larger sample size and a greater likelihood of reaching a definitive conclusion. The answer is yes, through a technique called *meta-analysis*.

Unfortunately, a complete description of this method is beyond the scope of this text. There are some restrictions on meta-analysis, for example, that the experiments whose p-values are to be combined should be comparable in nature.

Exercise 6.23: List all the respects in which you feel experiments ought be comparable in order that their p-values may be combined in a meta-analysis.

6.5   SUMMARY AND REVIEW

In this chapter, you learned the principles underlying the design and conduct of experiments and surveys. You learned how to cope with variation through controlling, blocking, measuring or randomizing with respect to all contributing factors. You learned the importance of giving a precise explicit formulation to your objectives and hypotheses. You learned a variety of techniques to ensure that your samples would be both random and representative of the population of interest. And you learned a variety of methods for determining the appropriate sample size.

You also learned that there is much more to statistics than can be presented within the confines of a single introductory text.

Exercise 6.24: A highly virulent disease is known to affect one in 5000 people. A new vaccine promises to cut this rate in half. Suppose we were to do an experiment in which we vaccinated a large number of people, half with an ineffective saline solution and half with the new vaccine. How many people would we need to vaccinate to ensure that the probability was 80% of detecting a vaccine as effective as this one purported to be while the risk of making a Type I error was no more than 5%? (*Hint:* See Section 4.2.1.)

There was good news and bad news when one of us participated in just such a series of clinical trials recently. The good news was that almost none of the subjects—control or vaccine treated—came down with the disease. The bad news was that with so few diseased individuals the trials were inconclusive.

Exercise 6.25: To compare teaching methods, 20 school children were randomly assigned to one of two groups. The following are the test results:

```
Conventional      85  79  80  70  61  85  98  80  86  75
New               90  98  73  74  84  81  98  90  82  88
```

Are the two teaching methods equivalent in result?

What sample size would be required to detect an improvement in scores of 5 units 90% of the time where our test is carried out at the 5% significance level?

Exercise 6.26: To compare teaching methods, 10 school children were divided into two groups. In one, they were first taught by conventional methods, tested and then taught related but different material by an entirely new approach. The second group were taught by the new approach first. The following are the test results:

```
Conventional      85  79  80  70  61  85  98  80  86  75
New               90  98  73  74  84  81  98  90  82  88
```

Are the two teaching methods equivalent in result?

What sample size would be required to detect an improvement in scores of five units 90% of the time? Again, the significance level for the hypothesis test is 5%.

Exercise 6.27: Make a list of all the italicized terms in this chapter. Provide a definition for each one along with an example.

Chapter 7

Guide to Entering, Editing, Saving, and Retrieving Large Quantities of Data Using R

7.1 CREATING AND EDITING A DATA FILE

Let's begin by reentering the class data from Chapter 1:

> ➢ height = c(141, 156.5, 162, 159, 157, 143.5, 154, 158, 140, 142, 150, 148.5, 138.5, 161, 153, 145, 147, 158.5, 160.5, 167.5, 155, 137)
> ➢ armspan = c(141, 156.5, 162, 159, 158, 143.5, 155.5, 160, 140, 142.5, 148, 148.5, 139, 160, 152.5, 142, 146.5, 159.5, 160.5, 164, 157, 137.5)

and put them together into a data frame we can add new data to:

> ➢ tdata=data.frame(height,armspan)

Now let's add some data

> ➢ classdata=edit(tdata)

When the edit window opens as shown in Figure 7.1, we click on the heading "var3" in column 3.

Next, we add entries to the column we've created as shown in Figure 7.2.

Introduction to Statistics Through Resampling Methods and R, Second Edition. Phillip I. Good.
© 2013 John Wiley & Sons, Inc. Published 2013 by John Wiley & Sons, Inc.

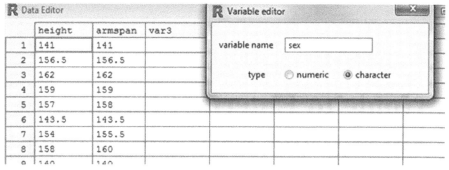

Figure 7.1   The R **edit**() window.

	height	armspan	sex	v
1	141	141	b	
2	156.5	156.5	g	
3	162	162	g	
4	159	159	g	
5	157	158	g	

Figure 7.2   Using the R Data Editor to add new entries to our data frame.

7.2   STORING AND RETRIEVING FILES FROM WITHIN R

- Create a directory for your R data files, for example, **dir.create("C:/Rdata")**
- Save your files, **save(classdata, file="C:/Rdata/classdat.dat")**.

You can save several files simultaneously as shown by the following command:

```
➢ save(classdata, ourdata, file="C:/Rdata/pak1.dat")
```

To retrieve your data set later on, type

```
➢ load("c:/Rdata/pak1.dat")
```

R and computer software in general is intolerant of errors. Notice that our quote marks are straight up and down and that our diagonal slashes are in the opposite direction of those used in Microsoft Windows®. When we load a file, we must use exactly the same spelling that we used when we saved it.

7.3 RETRIEVING DATA CREATED BY OTHER PROGRAMS

7.3.1 The Tabular Format

Suppose we have our data stored in a file called class.dat on a hard disk or CD, and that this file has the following form:

1. The first line of class.dat has a name for each variable in the data set.
2. Each additional line or record in class.dat has as its first item a row label followed by the values for each variable in the record, separated by blanks.

Here is an example:

	Name	Age	Gender	Height	ArmSpan
01	Pedro	11	B	167	163
02	Renee	11	G	163	162.5
03	Bill	10	B	153	152

We bring class.dat into our computer's memory and place it for use in an R frame with the command

```
> classdat = read.table("class.dat")
```

If class.dat was on a hard disk in the file folder c:\mywork or c:/mywork, we would store it for use in an R frame with the command,

```
> classdat = read.table("c:/mywork/class.dat")
```

Note: Use this form of referring to a file folder whether you are working in a Windows or UNIX environment. Working on a MAC use

```
classdat = read.table("./R/Good/Data/execcomp.txt")
```

This works from the current working directory of R (set in either R or the MAC GUI of R)

To use an absolute path:

```
lsat <- read.table("/Users/rob/Projects/R/R Books/
Good/Data/lsat.txt" )
```

7.3.2 Comma-Separated Values

If commas rather than blanks separate the data items in each record of your file, use the R command

```
> classdat = read.table("c:/mywork/class.dat", sep=",")
```

Before we may use the classdat variable names we must first attach the data frame with the command

```
➢ attach(classdat)
```

Then we may execute R commands, such as

```
➢ mean (Height)
➢ boxplot(ArmSpan)
```

Note that R commands are case sensitive; the following instructions won't work:

```
➢ mean (height)
➢ Mean(Height)
```

7.3.3 Data from Microsoft Excel

If you save in .csv format, then you can read in the data as described above. One word of caution, if you are saving column labels (and Excel may do this on its own), you need to use the command

```
classdat=read.table ("c:/rdata/class.dat",sep=",",
header=T)
```

7.3.4 Data from Minitab, SAS, SPSS, or Stata Data Files

There's an easy way and a hard way to make use of files created by these programs. The easy way is to export the data from these programs in .csv format and treat the result as you would files generated by Excel.

7.4 USING R TO DRAW A RANDOM SAMPLE

The easy part of using R to draw a sample is the part where the data we want is already in the computer memory and is stored in the vector "DATA." If we want 35 observations from DATA selected at random, we would just need to write

```
➢ sample (DATA, 35).
```

Now suppose we have our data stored in a file called class.dat on a hard disk or CD, and that this file has the following form:

1. The first line of class.dat has a name for each variable in the data set.
2. Each additional line or record in class.dat has as its first item a row label followed by the values for each variable in the record, separated by blanks.

Here is an example:

	Name	Age	Gender	Height	ArmSpan
01	Pedro	11	B	167	163
02	Renee	11	G	163	162.5
03	Bill	10	B	153	152

We could bring class.dat into our computer's memory and place it for use in an R frame with the command

```
> classdat = read.table("class.dat")
```

If class.dat was on a hard disk in the file folder \mywork or /mywork, we would store it for use in an R frame with the command,

```
>classdat = read.table("/mywork/class.dat")
```

(*Note*: Use this form of referring to a file folder no matter whether you are working in a Windows, UNIX or Mac environment.)

If commas rather than blanks separate the data items in each record of your file, use the R command

```
> classdat = read.table("/mywork/class.dat", sep=",")
```

Before we may use the classdat variable names, we must first attach the data frame with the command

```
> attach(classdat)
```

Then we may execute R commands, such as

```
> mean (Height)
> boxplot(ArmSpan)
```

Recall that R commands are case sensitive; the following instructions won't work:

```
> mean (height)
> boxplot(Armspan)
```

To select a sample of 10 random student records from classdat, use the commands

```
> nn = nrow(classdat)
> rr = sample(1:nn,10)
> samp = classdat[rr,]
```

or, put the three commands together into one.

```
samp = classdat [sample(1:nrow(classdat),10),]
```

R Programming Tip

Note that we use square brackets [ ] for vectors, and parenthesis ( ) for functions. Problems arise whenever we have an excess of opening symbols, such as [,{, or (over closing],},).

Now, let's put all we've learned together in one program:

```
➢ classdat = read.table("c:/mywork/class.dat")
➢ attach (classdat)
➢ samp = classdat [sample(1:nrow(classdat),10),]
➢ mean (Height)
```

[1] 161.2

```
➢ boxplot(ArmSpan)
```

Chapter 8

Analyzing Complex Experiments

In this chapter, you learn how to analyze a variety of different types of experimental data, including changes measured in percentages, measurements drawn from more than two populations, categorical data presented in the form of contingency tables, samples with unequal variances, and multiple endpoints.

8.1 CHANGES MEASURED IN PERCENTAGES

In the previous chapter, we learned how we could eliminate one component of variation by using each subject as its own control. But what if we are measuring weight gain or weight loss where the changes, typically, are best expressed as percentages rather than absolute values? A 250-pounder might shed 20 lb without anyone noticing; not so with an anorexic 105-pounder.

The obvious solution is to work not with before–after differences but with before/after ratios.

But what if the original observations are on growth processes—the size of a tumor or the size of a bacterial colony—and vary by several orders of magnitude? H.E. Renis of the Upjohn Company observed the following vaginal virus titres in mice 144 hours after inoculation with Herpes virus type II:

```
Saline controls        10,000, 3000, 2600, 2400, 1500.
Treated with antibiotic  9000, 1700, 1100,  360,     1.
```

In this experiment, the observed values vary from 1, which may be written as 10^0, to 10,000, which may be written as 10^4 or 10 times itself 4 times. With such wide variation, how can we possibly detect a treatment effect?

The trick employed by statisticians is to use the *logarithms* of the observations in the calculations rather than their original values. The logarithm or log of 10 is 1, the log of 10,000 written log(10,000) is 4. log(0.1) is −1. (Yes, the trick is simply to count the number of decimal places that follow the leading digit.)

Introduction to Statistics Through Resampling Methods and R, Second Edition. Phillip I. Good.
© 2013 John Wiley & Sons, Inc. Published 2013 by John Wiley & Sons, Inc.

Using logarithms with growth and percentage-change data has a second advantage. In some instances, it equalizes the variances of the observations or their ratios so that they all have the identical distribution up to a shift. Recall that equal variances are necessary if we are to apply any of the methods we learned for detecting differences in the means of populations.

Exercise 8.1: Was the antibiotic used by H.E. Renis effective in reducing viral growth? Hint: First convert all the observations to their logarithms using the R function **log**().

Exercise 8.2: While crop yield improved considerably this year on many of the plots treated with the new fertilizer, there were some notable exceptions. The recorded after/before ratios of yields on the various plots were as follows: 2, 4, 0.5, 1, 5.7, 7, 1.5, and 2.2. Is there a statistically significant improvement?

8.2 COMPARING MORE THAN TWO SAMPLES

The comparison of more than two samples is an easy generalization of the method we used for comparing two samples. As in Chapter 4, we want a test statistic that takes more or less random values when there are no differences among the populations from which the samples are taken but tends to be large when there are differences. Suppose we have taken samples of sizes $n_1, n_2, \ldots n_I$ from I distinct populations. Consider either of the statistics

$$F_2 = \sum_{i=1}^{I} n_i (\overline{X}_{i.} - \overline{X}_{..})^2,$$

that is, the weighted sum of the squared differences between the sample means and the grand means, where $\overline{X}_{i.}$ is the mean of the ith sample and $\overline{X}_{..}$ is the grand mean of all the observations, or

$$F_1 = \sum_{i=1}^{I} n_i |\overline{X}_{i.} - \overline{X}_{..}|,$$

that is, the weighted sum of the absolute values of the differences between the sample means and the grand means.

Recall from Chapter 1 that the symbol Σ stands for sum of, so that

$$\sum_{i=1}^{I} n_i (\overline{X}_{i.} - \overline{X}_{..})^2 = n_1 (\overline{X}_{1.} - \overline{X}_{..})^2 + n_2 (\overline{X}_{2.} - \overline{X}_{..})^2 + \cdots + n_I (\overline{X}_{I.} - \overline{X}_{..})^2.$$

If the means of the I populations are approximately the same, then changing the labels on the various observations will not make any difference as to the expected value of F_2 or F_1, as all the sample means will still have more or less the same magnitude. On the other hand, if the values in the first population are much larger than the values in the other populations, then our test statistic can only get smaller if we start rearranging the observations among the samples.

Exercise 8.3: To demonstrate that this is true, at least for samples of size 2, cut pairs of sticks, the members of each pair having the length of one of the observations, and use the pairs to fence in areas.

Our permutation test consists of rejecting the hypothesis of no difference among the populations when the original value of F_2 (or of F_1 should we decide to use it as our test statistic) is larger than all but a small fraction, say 5%, of the possible values obtained by rearranging labels.

8.2.1 Programming the Multi-Sample Comparison in R

To program the analysis of more than two populations in R, we need to program an R function. We also make use of some new R syntax to sum only those data values that satisfy a certain condition, sum(data[condition]), as in the expression term=mean(data[start : end]) which computes the mean of the data items beginning with data[start] and ending with data[end].

```
➤ F1=function(size,data){
➤ #size is a vector containing the sample sizes
➤ #data is a vector containing all the data in the
  #same order as the sample sizes
 +   stat=0
 +   start=1
 + grandMean = mean(data)
 + for (i in 1:length(size)){
 +    groupMean = mean(data[seq(from = start, length =
      size[i])])
 +    stat = stat + abs(groupMean - grandMean)
 +    start = start + size[i]
 +    }
 + return(stat)
 + }
```

We use this function repeatedly in the following R program:

```
➤ #One-way analysis of unordered data via a Monte
  #Carlo
➤ size = c(4,6,5,4)
➤ data = c(28,23,14.27, 33, 36,34, 29, 31, 34,
  18,21,20,22,24,11,14,11,16)
➤ #Insert the definition of function F1 here
➤ f1=F1(size,data)
➤ #Number N of simulations determines precision of
  #p-value
➤ N = 1600
➤ cnt = 0
```

```
➢ for (i in 1:N){
+ pdata = sample (data)
+ f1p=F1(size,pdata)
+ #counting number of rearrangements for which F1
  #greater than or equal to original
+ if (f1 <= f1p) cnt=cnt+1
+ }
➢ cnt/N
```

To test the program you've just created, generate 16 random values with the command

```
➢ data = rnorm(16, 2*rbinom(16, 1, .4))
```

The 16 values generated by this command come from a mixture of normal distributions $N(0,1)$ and $N(2,1)$. Set size $= c(4,4,3,5)$. Now do the following exercises.

Exercise 8.4: Run your program to see if it will detect differences in these four samples despite their all being drawn from the same population.

Exercise 8.5: Modify your program by adding the following command after you've generated the data

```
➢ data = data + c(2,2,2,2, 0,0,0,0, 0,0,0, 0,0,0,0,0)
```

Now test for differences among the populations.

Exercise 8.6: We saw in the preceding exercise that if the expected value of the first population was much larger than the expected values of the other populations, we would have a high probability of detecting the differences. Would the same be true if the mean of the second population was much higher than that of the first? Why?

Exercise 8.7: Modify your program by adding the following command after you've generated the data

```
➢ data = data + c(1,1,1,1, 0,0,0,0, -1.2,-1.2,-1.2,
  0 0,0,0,0)
```

Now test for differences among the populations.

***Exercise 8.8:** Test your command of R programming; write a function to compute F_2 and repeat the previous three exercises.

*8.2.2 Reusing Your R Functions

After you've created the function **F1**, you can save it to your hard disk with the R commands

```
➢ dir.create("C:/Rfuncs") #if the directory does not
  yet exist
➢ save(F1, file="C:/Rfuncs/pak1.Rfuncs")
```

When you need to recall this function for use at a later time, just enter the command

```
➢ load("C:/Rfuncs/pak1.Rfuncs ")
```

8.2.3 What Is the Alternative?

We saw in the preceding exercises that we can detect differences among several populations if the expected value of one population is much larger than the others or if the mean of one of the populations is higher and the mean of a second population is lower than the grand mean.

Suppose we represent the expectations of the various populations as follows: $EX_i = \mu + \delta_i$ where μ (pronounced mu) is the grand mean of all the populations and δ_i (pronounced delta sub i) represents the deviation of the expected value of the ith population from this grand mean. The sum of these deviations $\Sigma\delta_i = \delta_1 + \delta_2 + \ldots \delta_I = 0$. We will sometimes represent the individual observations in the form $X_{ij} = \mu + \delta_i + z_{ij}$, where z_{ij} is a random deviation with expected value 0 at each level of i. The permutation tests we describe in this section are applicable only if all the z_{ij} (termed the *residuals*) have the same distribution at each level of i.

One can show, though the mathematics is tedious, that the power of a test using the statistic F_2 is an increasing function of $\Sigma\delta_i^2$. The power of a test using the statistic F_1 is an increasing function of $\Sigma|\delta_i|$. The problem with these omnibus tests is that while they allow us to detect any of a large number of alternatives, they are not especially powerful for detecting a specific alternative. As we shall see in the next section, if we have some advance information that the alternative is an ordered dose response, for example, then we can develop a much more powerful statistical test specific to that alternative.

Exercise 8.9: Suppose a car manufacturer receives four sets of screws, each from a different supplier. Each set is a population. The mean of the first set is 4 mm, the second set 3.8 mm, the third set 4.1 mm and the fourth set 4.1 mm, also. What would be the values of μ, δ_1, δ_2, δ_3, and δ_4 be? What would be the value of $\Sigma|\delta_i|$?

8.2.4 Testing for a Dose Response or Other Ordered Alternative

D. Frank et al. (1978)* studied the increase in chromosome abnormalities and micronuclei as the doses of various compounds known to cause mutations were increased. Their object was to develop an inexpensive but sensitive biochemical test for mutagenicity that would be able to detect even marginal effects. The results of their experiment are reproduced in Table 8.1.

Table 8.1 Micronucleii in Polychromatophilic Erythrocytes and Chromosome Alterations in the Bone Marrow of Mice Treated with CY

Dose (mg/kg)	Number of Animals	Micronucelii per 200 cells	Breaks per 25 cells
0	4	0 0 0 0	0 1 1 2
5	5	1 1 1 4 5	0 1 2 3 5
20	4	0 0 0 4	3 5 7 7
80	5	2 3 5 11 20	6 7 8 9 9

* Evaluating drug-induced chromosome alterations, *Mutation Res.* 1978; 56: 311–317.

The best way to analyze these data is via the Pearson correlation applied not to the original doses but to their logarithms. To avoid problems with log(0), we add 1 to every dose before calculating its logarithm.

```
> logdose=log(c(rep(1,4),rep(6,5),rep(21,4),rep(81,5)))
> breaks= c(0,  1,  1,  2,   0,  1,  2,  3,  5,   3,  5,  7,
  7,   6,  7,  8,  9,  9)
> cor.test (logdose, breaks)
   Pearson's product-moment correlation
data: logdose and breaks
t = 6.8436, df = 16, p-value = 3.948e-06
alternative hypothesis: true correlation is not equal
to 0
95 percent confidence interval:
0.6641900 0.9480747
sample estimates:    cor    0.863345
```

A word of caution: If we use as the weights some function of the dose other than g[dose] = log[dose+1], we might observe a different result. Our choice of a test statistic must always make practical as well as statistical sense. In this instance, the use of the logarithm was suggested by a pharmacologist.

Exercise 8.10: Using the data for micronuclei, see if you can detect a significant dose effect.

Exercise 8.11: Aflatoxin is a common and undesirable contaminant of peanut butter. Are there significant differences in aflatoxin levels among the following brands?

Snoopy = c(2.5, 7.3, 1.1, 2.7, 5.5, 4.3)

Quick = c(2.5, 1.8, 3.6, 4.2 , 1.2 ,0.7)
MrsGood = c(3.3, 1.5, 0.4, 2.8, 2.2 ,1.0)

(*Hint*: What is the null hypothesis? What alternative or alternatives are of interest?)

Exercise 8.12: Does the amount of potash in the soil affect the strength of fibers made of cotton grown in that soil? Consider the data in the following table:

	Potash Level (lb/acre)				
	144	108	72	54	36
Breaking	7.46	7.17	7.76	8.14	7.63
Strength	7.68	7.57	7.73	8.15	8.00
	7.21	7.80	7.74	7.87	7.93

Exercise 8.13: Should you still perform this test if one or more of the observations in the previous table are missing?

8.3 EQUALIZING VARIABILITY

Suppose that to cut costs on our weight-loss experiment, we have each participant weigh his or herself. Some individuals will make very *precise* measurements, perhaps repeating the procedure three or four times to make sure they've performed the measurement correctly. Others, will say "close enough," and get the task done as quickly as possible. The problem with our present statistical methods is they treat each observation as if it were equally important. Ideally, we should give the least consideration to the most variable measurements and the greatest consideration to those that are least variable. The problem is that we seldom have any precise notion of what these variances are.

One possible solution is to put all results on a pass–fail or success–failure basis. This way, if the goal is to lose at least 7% of body weight, losses of 5 and 20% would offset each other, rather than a single 20% loss being treated as if it were equivalent to four losses of 5%. These pass-fail results would follow a binomial distribution and the appropriate method for testing binomial hypotheses could be applied.

The downside is the loss of information, but if our measurements are not equally precise, perhaps it is noise rather than useful information that we are discarding.

Here is a second example: An experimenter administered three drugs in random order to each of five recipients. He recorded their responses and now wishes to decide if there are significant differences among treatments. The problem is that the five subjects have quite different baselines. A partial solution would be to subtract an individual's baseline value from all subsequent observations made on that individual. But who is to say that an individual with a high baseline value will respond in the same way as an individual with a low baseline reading?

An alternate solution would be to treat each individual's readings as a *block* (see Section 6.2.3) and then combine the results. But then we run the risk that the results from an individual with unusually large responses might mask the responses of the others. Or, suppose the measurements actually had been made by five different experimenters using five different measuring devices in five different laboratories. Would it really be appropriate to combine them?

No, for sets of observations measured on different scales are not *exchangeable*. By converting the data to *ranks*, separately for each case, we are able to put all the observations on a common scale, and then combine the results.

When we replace observations by their ranks, we generally give the smallest value rank 1, the next smallest rank 2, and so forth. If there are N observations, the largest will have rank N. In creating Table 8.2b from Table 8.2a, we've made just such a transformation.

When three readings are made on each of five subjects, there are a total of $(3!)^5 = 7776$ possible rearrangements of labels within blocks (subjects). For a test of the null hypothesis against any and all alternatives using F_2 as our test statistic, as can be seen in Table 8.2b, only $2 \times 5 = 10$ of them, a handful of the total number of rearrangements, are as or more extreme than our original set of ranks.

Exercise 8.14: Suppose we discover that the measurement for Subject D for Treatment 1 was recorded incorrectly and should actually be 90. How would this affect the significance of the results depicted in Table 8.2?

Table 8.2a Original Observations

	A	B	C	D	E
Control	89.7	75	105	94	80
Treatment 1	86.2	74	95	98	79
Treatment 2	76.5	68	94	93	84

Table 8.2b Ranks

	A	B	C	D	E
Control	1	1	1	2	2
Treatment 1	2	2	2	1	3
Treatment 2	3	3	3	3	1

Exercise 8.15: Does fertilizer help improve crop yield when there is minimal sunlight? Here are the results of a recent experiment exactly as they were recorded. (Numbers in bushels.) No fertilizer: 5, 10, 8, 6, 9, 122. With fertilizer: 11, 18, 15, 14, 21, 14.

When to Use Ranks

1. When one or more extreme-valued observations are suspect.
2. When the methods used to make the measurements were not the same for each observation.
3. When the measurements were made on different scales (as in individual ratings of movies).

Many older textbooks advocate the use of tests based on ranks for a broad variety of applications. But *rank tests* are simply permutation tests applied to the ranks of observations rather than to their original values. The value of rank tests has diminished as a result of improvements in computer technology, and they should not be employed except in the three instances outlined above.

Ranks are readily obtained in R by use of the function **rank**().

8.4 CATEGORICAL DATA

We have shown in two examples how one may test for the independence of continuous measurements using a correlation coefficient. But what if our observations are categorical involving race or gender or some other categorical attribute?

Suppose on examining the cancer registry in a hospital, we uncover the following data that we put in the form of a 2×2 contingency table (Table 8.3). At issue is whether cancer survival is a function of the patient's sex.

Table 8.3   Cancer Survival as a Function of Sex

	Survived	Died	Total
Men	9	1	10
Women	4	10	14
Total	13	11	24

The 9 denotes the number of males who survived, the 1 denotes the number of males who died, and so forth. The four marginal totals or *marginals* are 10, 14, 13, and 11. The total number of men in the study is 10, while 14 denotes the total number of women, and so forth.

The marginals in this table are fixed because, indisputably, there are 11 dead bodies among the 24 persons in the study and 14 women. Suppose that before completing the table, we lost the subject IDs so that we could no longer identify which subject belonged in which category. Imagine you are given two sets of 24 labels. The first set has 14 labels with the word "woman" and 10 labels with the word "man." The second set of labels has 11 labels with the word "dead" and 12 labels with the word "alive." Under the null hypothesis that survival is independent of sex, you are allowed to distribute the two sets of labels to the subjects independently of one another. One label from each of the two sets per subject, please.

There are a total of $\binom{24}{10}$, 24 choose 10, ways you could hand out the gender labels "man" and "woman." $\binom{14}{10}\binom{10}{1}$ of these assignments result in tables that are as extreme as our original table (i.e., in which 90% of the men survive), and $\binom{14}{11}\binom{10}{0}$ in tables that are more extreme (100% of the men survive)—see Table 8.4a. In terms of the relative survival rates of the two sexes, Table 8.4a is more extreme than Table 8.3. Table 8.4b is less extreme.

Table 8.4a   Cancer Survival, Same Row and Column
Subtotals (marginals) as Table 8.3

	Survived	Died	Total
Men	10	0	10
Women	3	11	14
Total	13	11	24

Table 8.4b   Cancer Survival, Same Row and Column
Subtotals (marginals) as Table 8.3

	Survived	Died	Total
Men	8	2	10
Women	5	9	14
Total	13	11	24

This is a very small fraction of the total number of possible tables with the same marginals, less than 1%. To show this explicitly, we will need the aid of two R functions that we program ourselves: one that will compute factorials and one that will compute combinations. The first of these takes advantage of our knowledge that $n! = n(n - 1)!$ and uses a programming technique called *recursion*:

```
> fact=function(n){
+ if (n==1|n==0)1
+ else n*fact(n-1)
+ }
```

The preceding function assumes, without checking, that n is an integer. Let's try again and provide some solution messages in case we forget when we try to use this function a year from now:

```
> fact=function(n){
+ if(n!=trunc(n))return(print("argument must be a
  whole number"))
+ if(n<0) return(print ("argument must be nonnegative
  integer"))
+ else if(n==0|n==1)return(1)
+ else return(n*fact(n-1))
}
```

We then write $\binom{n}{t}$ in the form

```
> comb = function (n,t) fact(n)/(fact(t)*fact(n-t))
```

We want to compute

```
> (10*comb(14,10) + comb(14,11))/comb(24,10)
[1] 0.0056142594
```

From the result, 0.006, we conclude that a difference in survival rates of the two sexes at least as extreme as the difference we observed in our original table is very unlikely to have occurred by chance alone. We reject the hypothesis that the survival rates for the two sexes are the same and accept the alternative hypothesis that, in this instance at least, males are more likely to profit from treatment.

8.4.1 Making Decisions with R

In our factorial program, we used the if() command along with some logical operators. These include

== equal to

!= not equal to

as well as $<$, $<=$, $>$, and $>=$.

To combine conditions in a single if() use && meaning "and" or || meaning "or."

Warning: If you write if(X=2)A instead of if(X==2)A, you will set X=2, and the statements A will always be executed.

If **if(A)B** is followed by **else C**, B will be executed if A is true and C will not be executed. If A is false, then C will be executed, but B will not be.

Exercise 8.16: Write an R program to test your understanding of the **if()** and **else** commands. Test your program using several different data sets.

8.4.2 One-Sided Fisher's Exact Test

The preceding test is known as Fisher's exact test as it was first described by R.A. Fisher in 1935. Before we can write a program to perform this test, we need to consider the general case depicted in Table 8.5.

If the two attributes represented by the four categories are independent of one another, then each of the tables with the marginals n, m, and t is equally likely. If t is the smallest marginal, there are a total of $\binom{m+n}{t}$ possible tables. If $t - x$ is the value in the cell with the fewest observations, then

$$\sum_{k=0}^{t-x} \binom{m}{t-k}\binom{n}{k}$$

tables are as or more extreme than the one we observed.

Enter the actual cell frequencies

```
> data =c(f11, f12, f21, f22)
```

for example, data=c(9,1,4,10)

```
> data = array(data,dim=c(2,2)
> fisher.test(data)
```

Fisher's Exact Test for Count Data

```
data:  data
p-value = 0.004526
alternative hypothesis: true odds ratio is not equal to 1
95 percent confidence interval for the odds ratio:
1.748845 1073.050540
sample estimates: odds ratio 19.18092
```

Table 8.5   Relationships among Entries in a Contingency Table

	Category 1	Category 2	Total
Category A	$T - x$	X	T
Category B	$n - (t - x)$	$M - x$	$M + n - t$
Total	N	M	$m + n$

Exercise 8.17: a. Using the data in the following table, determine whether the new drug provides a statistically significant advantage over the old.

	Old Remedy	New Drug
Improved	6	10
No improvement	13	9

b. What must be true if this *p*-value is to be valid?

Exercise 8.18: What is the probability of observing Table 8.3 or a more extreme contingency table by chance alone?

Exercise 8.19: A physician has noticed that half his patients who suffer from sore throats get better within a week if they get plenty of bed rest. (At least they don't call him back to complain they aren't better.) He decides to do a more formal study and contacts each of 20 such patients during the first week after they've come to see him. What he learns surprises him. Twelve of his patients didn't get much of any bed rest or if they did go to bed on a Monday, they were back at work on a Tuesday. Of these noncompliant patients, six had no symptoms by the end of the week. The remaining eight patients all managed to get at least three days bed rest (some by only going to work half-days) and of these, six also had no symptoms by the end of the week. Does bed rest really make a difference?

8.4.3 The Two-Sided Test

In the example of the cancer registry, we tested the hypothesis that survival rates do not depend on sex against the alternative that men diagnosed with cancer are likely to live longer than women similarly diagnosed. We rejected the null hypothesis because only a small fraction of the possible tables were as extreme as the one we observed initially. This is an example of a one-tailed test. But is it the correct test? Is this really the alternative hypothesis we would have proposed if we had not already seen the data? Wouldn't we have been just as likely to reject the null hypothesis that men and women profit the same from treatment if we had observed a table of the following form?

Of course, we would! In determining the significance level in this example, we must perform a two-sided test and add together the total number of tables that lie in either of the two extremes or tails of the permutation distribution.

Unfortunately, it is not as obvious which tables should be included in the second tail. Is Table 8.6 as extreme as Table 8.3 in the sense that it favors an alternative more than the null hypothesis? One solution is simply to double the *p*-value we obtained for a one-tailed test. Alternately, we can define and use a test statistic as a basis of comparison. One commonly used measure is the Pearson χ^2 (chi-square) statistic defined for the 2×2 contingency table after eliminating terms that are invariant under permutations as $[x - tm/(m + n)]^2$. For Table 8.3, this statistic is 12.84; for Table 8.6, it is 29.34.

Table 8.6   Cancer Survival, Same Row and Column Subtotals (marginals) as Table 8.3

	Survived	Died	Total
Men	0	10	10
Women	13	1	14
Total	13	11	24

Table 8.7a   Cancer Survival, Same Row and Column Subtotals (marginals) as Table 8.3

	Survived	Died	Total
Men	1	9	10
Women	12	2	14
Total	13	11	24

Table 8.7b   Cancer Survival, Same Row and Column Subtotals (marginals) as Table 8.3

	Survived	Died	Total
Men	2	8	10
Women	11	3	14
Total	13	11	24

Exercise 8.20: Show that Table 8.7a is more extreme (in the sense of having a larger value of the chi-square statistic) than Table 8.3, but Table 8.7b is not.

8.4.4 Testing for Goodness of Fit

Recall that in Table 3.1, we compared the observed spatial distribution of superclusters of galaxies with a possible theoretical distribution.

```
➢ observed = c(7178,2012,370,38,1,1)
➢ theoretical= c(7095,2144,324,32,2,0)
➢ data = c(observed, theoretical)
➢ dim(data)=c(6,2)
➢ fisher.test(data)
      Fisher's Exact Test for Count Data
      data: data
      p-value = 0.05827
      alternative hypothesis: two.sided
```

8.4.5 Multinomial Tables

It is possible to extend the approach described in the previous sections to tables with multiple categories, such as Table 8.8, Table 8.9, and Table 8.10.

Table 8.8   3 × 3 Contingency Table

	Urban	Suburban	Rural
Republican			
Democrat			
Independent			

Table 8.9   Oral Lesions in Three Regions of India

Site of lesion	Kerala	Gujarat	Andhra
Labial mucosa	0	1	0
Buccal mucosa	8	1	8
Commissure	0	1	0
Gingiva	0	1	0
Hard palate	0	1	0
Soft palate	0	1	0
Tongue	0	1	0
Floor of mouth	1	0	1
Alveolar ridge	1	0	1

As in the preceding sections, we need only to determine the total number of tables with the same marginals as well as the number of tables that are as or more extreme than the table at hand. Two problems arise. First, what do we mean by more extreme? In Table 8.8, would a row that held one more case of "Full Recovery" be more extreme than a table that held two more cases of "Partial Recovery?" At least a half-dozen different statistics, including the Pearson $\chi 2$ statistic, have been suggested for use with tables like of Table 8.9 in which neither category is ordered.

Exercise 8.21: In a two-by-two contingency table, once we fix the marginals, we are only free to modify a single entry. In a three-by-three table, how many different entries are we free to change without changing the marginals? Suppose the table has R rows and C columns, how many different entries are we free to change without changing the marginals?

The second problem that arises lies in the computations, which are not a simple generalization of the program for the 2×2 case. Fortunately, R is able to perform the Fisher–Freeman–Halton test. Consider the data in Table 8.9:

Table 8.10   Plant Survival Under Varying Conditions

	No Fertilizer	Fertilizer B	Fertilizer C
Shade	1	2	1
Partial shade	2	3	4
Full sun	3	5	4

```
➢       Kerala = c(0,8,rep(0,5),1,1)
➢       Gujarat = c(rep(1,7),0,0)
➢       Andhra=Kerala
➢       data =c(Kerala,Gujarat,Andhra)
➢       nudata = array(data,dim=c(9,3))
➢       fisher.test(nudata)
    Fisher's Exact Test for Count Data
data: nudata
p-value = 0.0101
alternative hypothesis: two.sided
```

Exercise 8.22: Apply the Fisher–Freeman–Halton test to the data of Table 8.10.

8.5 MULTIVARIATE ANALYSIS

No competent physician would make a diagnosis on the basis of a single symptom. "Got a sore throat have you? Must be strep. Here take a pill." To the contrary, she is likely to ask you to open your mouth and say, "ah," take your temperature, listen to your chest and heartbeat, and perhaps take a throat culture. Similarly, a multivariate approach is far more likely to reveal differences brought about by treatment or environment than focusing on changes in a single endpoint.

Will a new drug help mitigate the effects of senility? To answer this question, we might compare treated and control patients on the basis of their scores on three quizzes: current events, arithmetic, and picture completion.

Will pretreatment of seeds increase the percentage of germination, shorten the time to harvest, and increase the yield?

Will a new fuel additive improves gas mileage and ride quality in both stop-and-go and highway situations?

In trying to predict whether a given insurance firm can avoid bankruptcy, should we measure the values and ratios of agents balances, total assets, stocks less cost, stocks less market value, expenses paid, and net premiums written?

In Section 8.5.2, we will learn how to combine the p-values obtained from separate but dependent single-variable (univariate) tests into an overall multivariate p-value.

8.5.1 Manipulating Multivariate Data in R

To take advantage of multivariate statistical procedures, we need to maintain all the observations on a single experimental unit as a single indivisible vector. If we are to do a permutation test, labels must be applied to and exchanged among these vectors and not among the individual observations. For example, if we are studying the effects of pretreatment on seeds, the data for each plot (the number of seeds that germinated, the time to the initial harvest, and the ultimate yield) must be treated as a single indivisible unit.

Data for the different variables may have been entered in separate vectors

```
➢ germ = c(11,9,8,4)
➢ time = c(5, 4.5, 6, 8)
```

Or, the observations on each experimental unit may have been entered separately.

```
➢ one = c(11,5)
➢ two = c (9,4.5)
➢ thr = c(8,6)
➢ four = c(4,8)
```

In the first case, we put the data in matrix form as follows

```
➢ temp = c(germ,time)
➢ data = matrix(temp,2,4, byrow=T)
➢ data
   [,1] [,2] [,3] [,4]
[1,]  11  9.0   8    4
[2,]   5  4.5   6    8
```

If the observations on each experimental unit have been entered separately, we would create a matrix this way;

```
➢ temp = c(one,two,thr, four)
➢ data = matrix (temp, 2,4)
➢ data
   [,1] [,2] [,3] [,4]
[1,]  11  9.0   8    4
[2,]   5  4.5   6    8
```

We can rank all the observations on a variable-by-variable basis with the commands

```
➢ R1 = rank(data[1,])
➢ R2 = rank(data[2,])
➢ rdata = matrix(c(R1,R2),2,4,byrow=T)
```

Here is a useful R function for generating random rearrangements of multivariate data:

```
➢ rearrange =function (data){
    n=length(data[1, ])
    m= length (data[ ,1])
    vect = c(1:n)
    rvect = sample (vect)
```

```
    new = data
    for (j in 1:n){
          ind = rvect [j]
          for (k in 1:m) new[k,j]= data [k,ind]
          }
return(new)
}
```

8.5.2 Hotelling's T^2

In the univariate (single-variable) case, Student's t and R's **t.test** serve for the comparison of the means of two populations and permutation tests are seldom necessary except for very small samples. Unlike Student's t, Hotelling's T^2, the multivariate equivalent of Student's t, is highly sensitive to departures from normality, and a distribution-free permutation approach to determining a p-value is recommended. To perform a permutation test with multivariate observations, we treat each vector of observations on an individual subject as a single indivisible entity. When we rearrange the treatment labels, we relabel on a subject-by-subject basis so that all observations on a single subject receive the same new treatment label.

Proceed in three steps:

1. Compute Hotelling's T^2 for the original observations.

2. Compute Hotelling's T^2 for each relabeling of the subjects.

3. Determine the percentage of relabelings that lead to values of the test statistic that are as or more extreme than the original value.

```
#This function for computing Hotelling's T² is due to
Nick Horton
hotelling = function(y1, y2) {
  #check the appropriate dimension
  k = ncol(y1)
  k2 = ncol(y2)
  if (k2!=k)
    stop("input matrices must have the same number of
    columns: y1 has ",
    k, " while y2 has ", k2)

  #calculate sample size and observed means
  n1 = nrow(y1)
  n2 = nrow(y2)
  ybar1 = apply(y1, 2, mean); ybar2 = apply(y2, 2, mean)
  diffbar = ybar1-ybar2

  #calculate the variance of the difference in means
  v = ((n1-1)*var(y1) + (n2-1)*var(y2))/(n1+n2-2)

  #calculate the test statistic and associated
  #quantities
  #%x% multiplies two matrices
  t2 = n1*n2*diffbar%*%solve(v)%*%diffbar/(n1+n2)
```

```
f = (n1+n2-k-1)*t2/((n1+n2-2)*k)
pvalue = 1-pf(f, k, n1+n2-k-1)

#return the list of results
 return(list(pvalue=pvalue, f=f, t2=t2, diff=diffbar))
}
#Compute the exact p-value
photell= function (samp1,samp2,fir,last,MC){
#fir=first of consecutive variables to be included in
#the analysis
#last=last of consecutive variables to be included in
#the analysis
#compute lengths of samples
d1=dim(samp1)[1]
d2=dim(samp2)[1]
d=d1+d2
samp=c(samp1,samp2)
#create dataframe to hold rearranged data
nsamp=samp
#compute test statistic for data
res=hotelling(samp[1:d1, fir:last],samp[(1+d1):d,fir:la
st])
stat0=as.numeric(res[[3]])
#run Monte Carlo simulation of p-value
cnt=1
for (j in 1:MC){
    nu=sample(lgth)
    for (k in 1:d)nsamp[nu[k],]=samp[k,]
    res=hotelling(nsamp[1:d1, fir:lst],nsamp[(1+d1):d,
    fir:last])
    if (as.numeric(res[[3]])>=stat0)cnt=cnt+1
    }
return (cnt)
}
print(c("pvalue=",cnt/MC), quote=FALSE)
```

*8.5.3   Pesarin–Fisher Omnibus Statistic

Let **Data** denote the original $N \times M$ matrix of multivariate observations. N, the number of rows in the matrix, corresponds to the number of experimental units. M, the number of columns in this matrix, corresponds to the number of different variables.

Our first task is to replace the matrix by a row vector T_0 of M univariate statistics. That is, calculating a column at a time, we replace that column with a single value—a t-statistic, a mean, or some other value.

To fix ideas, suppose we wish to compare two lots of fireworks, and have taken a sample from each. For each firework, we have recorded whether or not it exploded, how long it burned, and some measure of its brightness. The first column of the **Data** matrix consists of a series of ones corresponding to whether or not the firework for that row exploded. Such observations are best evaluated by the methods of Section 8.4, so we replace the column by a single number, that of the value of the Pearson chi-square statistic.

The times that are recorded in the second column have approximately an exponential distribution (see Section 4.4). So unless the observation is missing (as it would be if the firework did not explode), we take its logarithm, and replace the column by the mean of the observations in the first sample.

The brightness values in the third column have approximately a normal distribution, so we replace this column by the corresponding value of the t-statistic.

Our second task is to rearrange the observations in DATA between the two samples, using the methods of the previous section so as to be sure to keep all the observations on a given experimental unit together. Again, we compute a row vector of univariate statistics **T**. We repeat this resampling process K times. If there really isn't any difference between the two lots of fireworks, then each of the vectors $\mathbf{T_k}$ obtained after rearranging the data as well as the original vector $\mathbf{T_0}$ are equally likely.

We would like to combine these values but have the same problem we encountered and resolved in Section 8.3: the p-values have all been measured on different scales. The solution as it was in Section 8.3 is to replace the $K + 1$ different values of each univariate test statistic by their ranks. We rank the $K + 1$ values of the χ^2 statistic separately from the $K + 1$ values of the t-statistic, and so forth, so that we have m sets of ranks ranging from 1 to $K + 1$.

The ranks should be related to the extent to which the statistic favors the alternative. For example, if large values of the t-statistic are to be expected when the null hypothesis is false, then the largest value of this statistic should receive rank $K + 1$.

Our third task is to combine the ranks of the M individual univariate tests using Fisher's omnibus statistic

$$U_i = -\sum_{m=1}^{M} \log\left[\frac{K+0.5-R_{im}}{K+1}\right]; i = 1, \ldots K+1.$$

Our fourth and final task is to determine the p-value associated with this statistic. We require three sets of computations. To simplify writing them down, we will make repeated use of an indicator function I. $I[y] = 1$ if its argument y is true, $I[y] = 0$ otherwise. For example, $I[T>3]$ is 1 if $T > 3$ and is 0 otherwise.

1. Determine from the individual distributions (parametric or permutation) the marginal significance level of each of the single-variable statistics for the original nonrearranged observations,

$$p_m = \frac{0.5 + \sum_{k=1}^{K} I[T_{km} \geq T_{0m}]}{K+1}, m = 1, \ldots, M.$$

2. Combine these values into a single statistic

$$U_0 = -\sum_{m=1}^{M} \log[p_m] = -\sum_{m=1}^{M} \log\left[\frac{K + 0.5 - R_{0m}}{K + 1}\right].$$

Note that R_{ok} can take any value in the range 1 to $K + 1$.

3. Determine the significance level of the combined test,

$$p = \frac{0.5 + \sum_{k=1}^{K} I[U_k \geq U_0]}{K + 1}.$$

8.6 R PROGRAMMING GUIDELINES

To perform the Pesarin–Fisher omnibus multivariate test, you will need to write a program in R. The program is a long one and it is essential that you proceed in a systematic fashion. Here are some guidelines.

- Envision the program in sections. For example:
  1. Enter data and store in matrix form.
  2. Compute univariate statistics.
  3. Repeatedly resample from data and compute univariate statistics.
  4. Convert univariate statistics to ranks.
  5. Compute the Pesarin–Fisher statistic and determine its p-value.
- Write general, but test simple.
  Use global constants for sample sizes and number of resamples so you can make changes at a single location.
- Program and test your program a section at a time; use a variety of inputs
- Insert a comment before each line of your program describing what that line does and why. This last suggestion may seem silly and unnecessary, but suppose you need to use and modify your program a month or a year from today. Will you still know what each line of your program is supposed to do?
- Define and use functions. For example, by defining a function to compute all your univariate statistics (each of which requires its own function), you can use that function first to compute the univariate statistics for the data with its original labeling and then for each rearrangement. Think first, program second. Organize your program first on paper and then transfer it section by section to the computer, filling in the line-by-line details and testing as you go. Save your work.

Anticipate and program for exceptions. For example:

```
if (n<0 || k<0 || (n-trunc(n))!=0 || (k-trunc(k))!=0)
stop("n and k must be positive integers")else
fact(n)/(fact(k)*fact(n-k))
```

Testing and debugging a program also requires that you proceed in a systematic fashion:

- Use artificial data sets to begin with. Use the smallest possible number of samples, variables, and experimental units, for example, two samples, each with two experimental units, and two observations on each experimental unit. Use distinct samples, such as (1,2) and (0,0).
- Test a section at a time. When testing the section in which you resample, test with $N = 1$ and then $N = 2$ before increasing N to a more practical value. If the results don't make sense, print out intermediate values. For example, after you enter the data,
  ○ data = matrix (temp, 2,4);

  display it
  ○ data.

Once you are sure your program works correctly, you can remove these redundant commands.

- Vary your test data a step at a time. For example, use three observations per sample instead of two. Next, use unequal sample sizes. Then, use three observations (variables) per experimental unit instead of three.
- Complete your tests using real data.

Exercise 8.23: You are studying a new tranquilizer that you hope will minimize the effects of stress. The peak effects of stress manifest themselves between 5 and 10 minutes after the stressful incident, depending on the individual. To be on the safe side, you've made observations at both the 5- and 10-minute marks.

Subject	Prestress	5-minute	10-minute	Treatment
A	9.3	11.7	10.5	Brand A
B	8.4	10.0	10.5	Brand A
C	7.8	10.4	9.0	Brand A
D	7.5	9.2	9.0	New drug
E	8.9	9.5	10.2	New drug
F	8.3	9.5	9.5	New drug

How would you correct for the prestress readings? Is this a univariate or a multivariate problem? List possible univariate and multivariate test statistics. Perform the permutation tests and compare the results.

Exercise 8.24: To help understand the sources of insolvency, J. S. Trieschman and G. E. Pinches (1973)* compared the financial ratios of solvent and financially distressed insurance firms. A partial listing of their findings is included in the following table. Are the differences statistically significant? Be sure to state the specific hypotheses and alternative hypotheses you are testing. Will you use Student's t as your univariate test statistic each time? Or the difference in mean values? Or simply the sum of the observations in the first sample?

* A multivariate model for predicting financially distressed P-L insurers, *J. Risk Insurance.* 1973; 40: 327–338.

Solvent companies					Insolvent companies			
	V1	V2	V3	V4		V1	V2	V3	V4
1	0.056	0.398	1.138	0.109	9	0.059	1.168	1.145	0.732
2	0.064	0.757	1.005	0.085	10	0.054	0.699	1.052	0.052
3	0.033	0.851	1.002	0.118	11	0.168	0.845	0.997	0.093
4	0.025	0.895	0.999	0.057	12	0.057	0.592	0	0.057
5	0.050	0.928	1.206	0.191	13	0.337	0.898	1.033	0.088
6	0.060	1.581	1.008	0.146	14	0.230	1	1.157	0.088
7	0.015	0.382	1.002	0.141	15	0.107	0.925	0.984	0.247
8	0.079	0.979	0.996	0.192	16	0.193	1.120	1.058	0.502

V1 = (agents balances)/(total assets); V2 = (stocks − cost)/(stocks − market value); V3 = (bonds − cost)/(bonds − market value); V4 = (expenses paid)/(net premiums written).

Exercise 8.25: You wish to test whether a new fuel additive improves gas mileage and ride quality in both stop-and-go and highway situations. Taking 12 vehicles, you run them first on a highway-style track and record the gas mileage and driver's comments. You then repeat on a stop-and-go track. You empty the fuel tanks and refill, this time including the additive, and again run the vehicles on the two tracks.

The following data were supplied in part by the Stata Corporation. Use Pesarin's combination method to test whether the additive affects either gas mileage or ride quality.

Id	bmpg1	ampg	rqi1	bmpg2	ampg2	rqi2
1	20	24	0	19	23.5	1
2	23	25	0	22	24.5	1
3	21	21	1	20	20.5	0
4	25	22	0	24	20.5	−1
5	18	23	1	17	22.5	1
6	17	18	−1	16	16.5	−1
7	18	17	0	17	16.5	0
8	24	28	1	23	27.5	0
9	20	24	0	19	23.5	1
10	24	27	0	22	25.5	0
11	23	21	0	22	20.5	0
12	19	23	1	18	22.5	1

bmpg1, track 1 before additive; ampg1, track 1 after additive; rqi1, ride quality improvement track 1; bmpg2, track 2 before additive; ampg2, track 2 after additive; rqi2, ride quality improvement track 2.

Exercise 8.26: The following data are based on samples taken from swimming areas off the coast of Milazzo (Italy) from April through September 1998. Included in this data set are levels of total coliform, fecal coliform, streptococci, oxygen, and temperature.

- Are there significant differences in each of these variables from month to month?
- Are there significant multivariate differences from month to month in terms of the Fisher–Pesarin omnibus statistic?
- Was the use of a multivariate approach of value in this example?

```
Month = c(4,  5,  6,  7,  8,  9,  4,  5,  6,  7,  8,  9,  4,  5,
6,  7,  8,  9,  4,  5,  6,  7,  8,  9,  4,  5,  6,  7,  8,  9,  4,
5,  6,  7,  8,  9,  4,  5,  6,  7,  8,  9,  4,  5,  6,  7,  8,  9,
4,  5,  6,  7,  8,  9,  4,  5,  6,  7,  8,  9,  4,  5,  6,  7,  8,
9,  4,  5,  6,  7,  8,  9,  4,  5,  6,  7,  8,  9,  4,  5,  6,  7,
8,  9)
Temp = c(14,  17,  24,  21,  22,  20,  14,  17,  24,  21,  23,
22,  14,  17,  25,  21,  21,  22,  14,  17,  25,  21,  25,  20,
14,  17,  25,  21,  21,  19,  14,  17,  25,  21,  25,  19,  14,
16,  25,  21,  25,  19,  15,  19,  18,  21,  25,  19,  15,  19,
18,  20,  22,  17,  15,  19,  18,  20,  23,  18,  15,  17,  18,
20,  25,  17,  15,  17,  18,  20,  25,  19,  15,  17,  19,  20,
25,  18,  15,  18,  19,  21,  24,  19)
TotColi = c(30,  22,  16,  18,  32,  40,  50,  34,  32,  32,
34,  18,  16,  19,  65,  54,  32,  59,  45,  27,  88,  32,  78,
45,  68,  14,  54,  22,  25,  32,  22,  17,  87,  17,  46,  23,
10,  19,  38,  22,  12,  26,  8,  8,  11,  19,  45,  78,  6,  9,
87,  6,  23,  28,  0,  0,  43,  8,  23,  19,  0,  5,  28,  19,
14,  32,  12,  17,  33,  21,  18,  5,  22,  13,  19,  27,  30,
28,  16,  6,  21,  27,  58,  45)
FecColi = c(16,  8,  8,  11,  11,  21,  34,  11,  11,  7,  11,
6,  8,  6,  35,  18,  18,  21,  13,  9,  32,  11,  29,  11,  28,
7,  12,  7,  12,  9,  10,  3,  43,  5,  12,  14,  4,  9,  8,  10,
4,  12,  0,  4,  7,  5,  12,  26,  0,  3,  32,  0,  8,  12,  0,  0,
21,  0,  7,  8,  0,  0,  17,  4,  0,  14,  0,  0,  11,  7,  6,  0,
8,  0,  6,  4,  5,  10,  14,  3,  8,  12,  11,  27)
Strep = c(8,  4,  3,  5,  6,  9,  18,  7,  6,  3,  5,  2,  6,  0,
11,  2,  12,  7,  9,  3,  8,  6,  9,  4,  8,  2,  7,  3,  5,  3,
4,  0,  16,  0,  6,  5,  0,  0,  2,  4,  0,  4,  0,  0,  3,  0,  5,
14,  0,  0,  11,  0,  2,  4,  0,  0,  3,  0,  2,  2,  0,  0,  7,
2,  0,  5,  0,  0,  7,  4,  3,  0,  2,  0,  2,  2,  3,  5,  8,  1,
2,  4,  5,  7)
O2 = c(95.64,  102.09,  104.76,  106.98,  102.6,  109.15,
96.12,  111.98,  100.67,  103.87,  107.57,  106.55,  89.21,
100.65,  100.54,  102.98,  98,  106.86,  98.17,  100.98,
99.78,  100.87,  97.25,  97.78,  99.24,  104.32,  101.21,
102.73,  99.17,  104.88,  97.13,  102.43,  99.87,  100.89,
99.43,  99.5,  99.07,  105.32,  102.89,  102.67,  106.04,
106.67,  98.14,  100.65,  103.98,  100.34,  98.27,  105.69,
96.22,  102.87,  103.98,  102.76,  107.54,  104.13,  98.74,
101.12,  104.98,  101.43,  106.42,  107.99,  95.89,
```

```
104.87, 104.98, 100.89, 109.39, 98.17, 99.14, 103.87,
103.87, 102.89, 108.78, 107.73, 97.34, 105.32,
101.87, 100.78, 98.21, 97.66, 96.22, 22, 99.78,
101.54, 100.53, 109.86)
```

8.7   SUMMARY AND REVIEW

In this chapter, you learned to analyze a variety of different types of experimental data. You learned to convert your data to logarithms when changes would be measured in percentages or when analyzing data from dividing populations. You learned to convert your data to ranks when observations were measured on different scales or when you wanted to minimize the importance of extreme observations.

You learned to specify in advance of examining your data whether your alternative hypotheses of interest were one-sided or two-sided, ordered or unordered.

You learned how to compare binomial populations and to analyze 2×2 contingency tables. You learned how to combine the results of multiple endpoints via Hotelling's T^2 and the Pesarin–Fisher omnibus multivariate test.

You were introduced to the built-in R math functions log, exp, rank, abs, and trunc , used R functions cor.test and fisher.test and created several new R functions of your own with the aid of if . . . else commands and return(list()). You learned how to store data in matrix form using the R functions array, dim, and matrix, multiply matricies using %*%, and extract the marginal of matricies using the apply function. You learned how to print results along with techniques for the systematic development and testing of lengthy programs.

Exercise 8.27: Write definitions for all italicized words in this chapter.

Exercise 8.28: Many surveys employ a nine-point Likert scale to measure respondent's attitudes where a value of "1" means the respondent definitely disagrees with a statement and a value of "9" means they definitely agree. Suppose you have collected the views of Republicans, Democrats, and Independents in just such a form. How would you analyze the results?

Chapter 9

Developing Models

A recent help wanted ad touted openings with a sportsbook; they were looking for statisticians who could help them develop real-time software for forecasting the outcomes of cricket matches. In this chapter, you will learn valuable techniques with which to develop forecasts and classification schemes. These techniques have been used to forecast parts s4ales by the Honda Motors Company and epidemics at naval training centers, to develop criteria for retention of marine recruits, optimal tariffs for Federal Express, and multitiered pricing plans for Delta Airlines. And these are just examples in which I've been personally involved!

9.1 MODELS

A model in statistics is simply a way of expressing a quantitative relationship between one variable, usually referred to as the *dependent variable,* and one or more other variables, often referred to as the *predictors.* We began our text with a reference to Boyle's law for the behavior of perfect gases, $V = KT/P$. In this version of Boyle's law, V (the volume of the gas) is the dependent variable; T (the temperature of the gas) and P (the pressure exerted on and by the gas) are the predictors; and K (known as Boyle's constant) is the *coefficient* of the ratio T/P.

An even more familiar relationship is that between the distance S traveled in t hours and the velocity V of the vehicle in which we are traveling: $S = Vt$. Here, S is the dependent variable, and V and t are predictors. If we travel at a velocity of 60 mph for 3 hours, we can plot the distance we travel over time, using the following R code:

```
> Time = c(0.5,1,1.5,2, 2.5, 3)
> S = 60*Time
> plot(S, xlab="Time")
```

I attempted to drive at 60 mph on a nearby highway past where a truck had recently overturned with the following results:

Introduction to Statistics Through Resampling Methods and R, Second Edition. Phillip I. Good.
© 2013 John Wiley & Sons, Inc. Published 2013 by John Wiley & Sons, Inc.

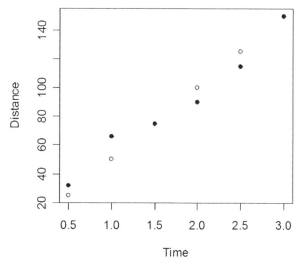

Figure 9.1   Distance expected at 50 mph versus distance observed.

```
> SReal = c(32,66,75,90,115,150)
> plot(Time,S, ylab="Distance")
> points(Time,SReal,pch=19)
```

As you can see, the reality was quite different from the projected result. My average velocity over the 3-hour period was equal to distance traveled divided by time or $150/3 = 50$ mph. That is, $S_i = 50*\text{Time}_i + z_i$, where the $\{z_i\}$ are random deviations from the expected distance.

Let's compare the new model with the reality:

```
> S = 50*Time
> SReal = c(32,66,75,90,115,150)
> plot(Time,S, ylab="Distance")
> points(Time,SReal,pch=19)
```

Our new model, depicted in Figure 9.1, is a much better fit, agreed?

9.1.1 Why Build Models?

We develop models for at least three different purposes. First, as the terms "predictors" suggests, models can be used for *prediction*. A manufacturer of automobile parts will want to predict part sales several months in advance to ensure its dealers have the necessary parts on hand. Too few parts in stock will reduce profits; too many may necessitate interim borrowing. So entire departments are hard at work trying to come up with the needed formula.

At one time, I was part of just such a study team working with Honda Motors. We soon realized that the primary predictor of part sales was the weather. Snow, sleet, and freezing rain sent sales skyrocketing. Unfortunately, predicting the weather is as or more difficult than predicting part sales.

Second, models can be used to develop additional insight into cause and effect relationships. At one time, it was assumed that the growth of the welfare case load L was a simple function of time t, so that $L = ct$, where the growth rate c was a function population size. Throughout the 1960s, in state after state, the constant c constantly had to be adjusted upward if this model were to fit the data. An alternative and better fitting model proved to be $L = ct + dt^2$, an equation often used in modeling the growth of an epidemic. As it proved, the basis for the new second-order model was the same as it was for an epidemic: Welfare recipients were spreading the news of welfare availability to others who had not yet taken advantage of the program much as diseased individuals might spread an infection.

Boyle's law seems to fit the data in the sense that if we measure both the pressure and volume of gases at various temperatures, we find that a plot of pressure times volume versus temperature yields a straight line. Or if we fix the volume, say by confining all the gas in a chamber of fixed size with a piston on top to keep the gas from escaping, a plot of the pressure exerted on the piston against the temperature of the gas yields a straight line.

Observations such as these both suggested and confirmed what is known today as kinetic molecular theory.

A third use for models is in *classification*. At first glance, the problem of classification might seem quite similar to that of prediction. Instead of predicting that Y would be 5 or 6 or even 6.5, we need only predict that Y will be greater or less than 6, for example. A better example might be assigning a mushroom as one of the following classifications: (1) edible and tasty, (2) poisonous, and (3) boring. The loss functions for prediction and classification are quite different. The loss connected with predicting y_p when the observed value is y_o is usually a monotone increasing function of the difference between the two. By contrast, the loss function connected with a classification problem has jumps, as in Table 9.1.

Not surprisingly, different modeling methods have developed to meet the different purposes. In the balance of the chapter, we shall consider three primary modeling methods: *decision trees*, which may be used both for classification and prediction, *linear regression*, whose objective is to predict the expected value of a given dependent variable, and *quantile regression*, whose objective is to predict one or more quantiles of the dependent variable's distribution.

Table 9.1   Classification Loss Matrix for Mushroom Example

	Decision		
	Edible and tasty	Poisonous	Boring
Edible and tasty	Great meal	Missed opportunity	Missed opportunity
Poisonous	Death	Nothing	Possibility of death?
Boring	So-so meal	Nothing	Nothing

9.1.2 Caveats

The modeling techniques that you learn in this chapter may seem impressive—they require extensive calculations that only a computer can do—so that I feel it necessary to issue three warnings.

- You cannot use the same data both to formulate a model and to test it. It must be independently validated.
- A cause-and-effect basis is required for every model, just as Molecular Theory serves as the casual basis for Boyle's law.
- Don't let your software do your thinking for you. Just because a model fits the data does not mean that it is appropriate or correct. Again, it must be independently validated and have a cause and effect basis.

You may have heard that a black cat crossing your path will bring bad luck. Don't step in front of a moving vehicle to avoid that black cat unless you have some causal basis for believing that black cats can affect your luck. (And why not white cats or tortoise shell?) I avoid cats myself because cats lick themselves and shed their fur; when I breathe cat hairs, the traces of saliva on the cat fur trigger an allergic reaction that results in the blood vessels in my nose dilating. Now, that is a causal connection.

9.2 CLASSIFICATION AND REGRESSION TREES

If you're a mycologist, a botanist, a herpetologist or simply a nature lover you may already have made use of some sort of a key. For example,

1. Leaves simple?
    1. Leaves needle-shaped?
        a. Leaves in clusters of 2 to many?
            i. Leaves in clusters of 2 to 5, sheathed, persistent for several years?

The R **rpart()** procedure attempts to construct a similar classification scheme from the data you provide. You'll need to install a new package of R functions in order to do the necessary calculations. The easiest way to do this is get connected to the Internet and then type

```
> install.packages("rpart")
```

The installation which includes downloading, unzipping, and integrating the new routines is then done automatically. Installation needs to be done once and once only. But each time, you'll need to load the decision-tree building functions into computer memory before you can us them by typing

```
> library(rpart)
```

9.2.1 Example: Consumer Survey

In order to optimize an advertising campaign for a new model of automobile by directing the campaign toward the best potential customers, a study of consumers' attitudes, interests, and opinions was commissioned. The questionnaire consisted of a number of statements covering a variety of dimensions, including consumers' attitudes towards risk, foreign-made products, product styling, spending habits, emissions, pollution, self-image, and family. The final question concerned the potential customer's attitude toward purchasing the product itself. All responses were tabulated on a nine-point Likert scale. The data may be accessed at http://statcourse.com/Data/ConsumerSurvey.htm.

Figure 9.2 is an example of a decisions tree we constructed using the data from the consumer survey and the R commands

```
➤ catt.tr=rpart(Purchase ~Fashion + Ozone + Gamble)
➤ plot(catt.tr, uniform=TRUE, compress=TRUE);
  text(catt.tr, use.n=TRUE)
```

Note that some of the decisions, that is, the branch points in the tree shown in the figure, seem to be made on the basis of a single predictor—Gamble; others utilized the values of two predictors—Gamble and Fashion; and still others utilized the values of Gamble, Fashion, and Ozone, the equivalent of three separate linear regression models.

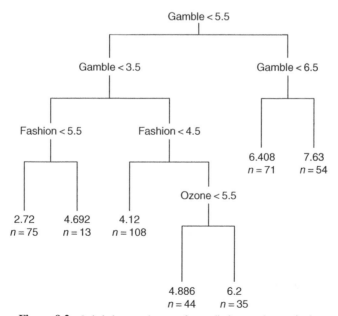

Figure 9.2   Labeled regression tree for predicting purchase attitude.

Exercise 9.1: Show that the R function **rpart()** only makes use of variables it considers important by constructing a regression tree based on the model

$$Purchase \sim Fashion + Ozone + Pollution + Coupons$$
$$+ Gamble + Today + Work + UPM.$$

Exercise 9.2: Apply the computer-aided regression tree (CART) method to the Milazzo data of the previous chapter in Exercise 8.24 to develop a prediction scheme for coliform levels in bathing water based on the month, oxygen level, and temperature.

The output from our example and from Exercises 9.1 and 9.2 are *regression trees*, that is, at the end of each *terminal node* is a numerical value. While this might be satisfactory in an analysis of our Milazzo coliform data, it is misleading in the case of our consumer survey, where our objective is to pinpoint the most likely sales prospects.

Suppose instead that we begin by grouping our customer attitudes into categories. Purchase attitudes of 1, 2, or 3 indicate low interest, 4, 5, and 6 indicate medium interest, 7, 8, and 9 indicate high interest. To convert them using R, we enter

```
➢ Pur.cat = cut(Purchase,breaks=c(0,3,6,9),labels=c(
  "lo", "med", "hi"))
```

Now, we can construct the classification tree depicted in Figure 9.3.

```
➢ catt.tr=rpart(Pur.cat~Fashion + Ozone + Gamble)
➢ plot(catt.tr, uniform=TRUE, compress=TRUE);
  text(catt.tr, use.n=TRUE)
```

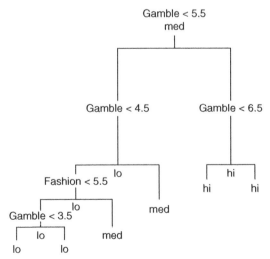

Figure 9.3 Labeled classification tree for predicting prospect attitude.

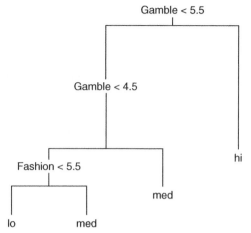

Figure 9.4   Pruned classification tree for predicting prospect attitude.

Many of the branches of this tree appear redundant. If we have already classified a prospect, there is no point in additional questions. We can remove the excess branches and create a display similar to Figure 9.4 with the following code:

```
➤ tmp=prune(catt.tr, cp=0.02)
➤ plot(tmp); text(tmp)
```

These classifications are actually quite tentative; typing the command

```
➤ print(tmp)
```

will display the nodes in this tree along with the proportions of correct and incorrect assignments.

```
n = 400
node), split, n, loss, yval, (yprob)
      * denotes terminal node
1) root 400 264 med (0.3375000 0.3400000 0.3225000)
  2) Gamble< 5.5 275 148 lo (0.4618182 0.3600000
     0.1781818)
    4) Gamble< 4.5 192 83 lo (0.5677083 0.2708333
       0.1614583)
      8) Fashion< 5.5 160 57 lo (0.6437500 0.2375000
         0.1187500) *
      9) Fashion>=5.5 32 18 med (0.1875000 0.4375000
         0.3750000) *
    5) Gamble>=4.5 83 36 med (0.2168675 0.5662651
       0.2168675) *
  3) Gamble>=5.5 125 45 hi (0.0640000 0.2960000
     0.6400000) *
```

The three fractional values within parentheses refer to the proportions in the data set that were assigned to the "low," "med," and "hi" categories, respectively. For example, in the case when Gamble > 5.5, 64% are assigned correctly by the classification scheme, while just over 6% who are actually poor prospects (with scores of 3 or less) are assigned in error to the high category.

> **Exercise 9.3:** Goodness of fit does not guarantee predictive accuracy. To see this for yourself, divide the consumer survey data in half. Use the first part for deriving the tree, and the second half to validate the trees use for prediction.
>
> a. Create a classification tree for Purchase as a function of **all** the variables in the consumer survey. Treat each Purchase level as a single category—this can be done in a single step by typing Pur.fact = factor(Purchase). Prune the tree if necessary.
> b. Validate the tree you built using the predict() function located in the tree library.

9.2.2 How Trees Are Grown

Classification and regression tree builders like **rpart**() make binary splits based on either yes/no (Sex = Male?) or less than/greater than (Fashion < 5.5) criteria. The tree is built a level at a time beginning at the root. At each level, the computationally intensive algorithm used by **rpart**() examines each of the possible predictors and chooses the predictor and the split value (if it is continuous) that best separates the data into more homogenous subsets.

> ➤ This method is illustrated in Figures 9.5a,b and 9.6 using the iris data set at http://statcourse. com/research/Iris.csv. The objective in this instance is to assign an iris to one of three species based on the length and width of its petals and its sepals.

The best fitting decision tree is

> ➤ rpart(Class~SepLth + PetLth + SepWth + PetWth)
> node), split, n, loss, yval, (yprob)
>   * denotes terminal node
> 1) root 150 100 setosa (0.33333333 0.33333333
>   0.33333333)
>   2) PetLth< 2.45 50   0 setosa (1.00000000 0.00000000
>   0.00000000) *
>   3) PetLth>=2.45 100 50 versicolor (0.00000000
>   0.50000000 0.50000000)
>   6) PetWth< 1.75 54   5 versicolor (0.00000000
>   0.90740741 0.09259259) *
>   7) PetWth>=1.75 46   1 virginica (0.00000000
>   0.02173913 0.97826087) *

in which sepal length and width play no part.

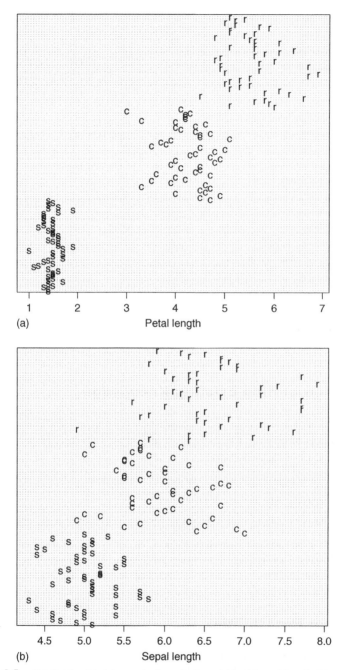

Figure 9.5   Subdividing iris into species by a single variable: (a) petal length, (b) sepal length.

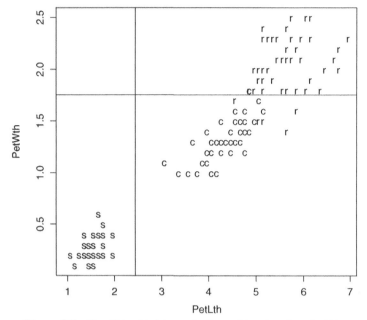

Figure 9.6 Classifying iris into species by petal length and petal width.

9.2.3 Incorporating Existing Knowledge

One can and should incorporate one's knowledge of prior probabilities and relative costs into models built with a classification and regression tree. In particular, one can account for all of the following:

1. The prior probabilities of the various classifications.
2. The relative costs of misclassification.
3. The relative costs of data collection for the individual predictors.

When decision trees are used for the purpose of analyzing biomedical images—sonograms, X-rays, EEGs, and PET scans, all three of these factors must be taken into consideration.

9.2.4 Prior Probabilities

Although equal numbers of the three species of Iris were to be found in the sample studied by Anderson, it is highly unlikely that you would find such a distribution in an area near where you live. Specifying that the setosa are the most plentiful yields a quite different decision tree. In the following R code, we specify that the setosa occur 60% of the time:

```
➤ rpart(Class~SepLth + PetLth + SepWth + PetWth, parms=
  list(prior=c(.20,.20,.60)))
  1) root 150 60.0 virginica (0.20000000 0.20000000
       0.60000000)
    2) PetLth< 4.75 95 28.2 setosa (0.51546392
         0.45360825 0.03092784)
      4) PetLth< 2.45 50 0.0 setosa (1.00000000
           0.00000000 0.00000000) *
      5) PetLth>=2.45 45 1.8 versicolor (0.00000000
           0.93617021 0.06382979) *
    3) PetLth>=4.75 55 3.6 virginica (0.00000000
         0.03921569 0.96078431) *
```

9.2.5 Misclassification Costs

Suppose instead of three species of Iris, we had data on three species of mushroom: the *Amanita calyptroderma* (edible and choice), *Amanita citrine* (probably edible), and the *Aminita verna* (more commonly known as the Destroying Angel). Table 9.1 tells us that the losses from misclassification vary from a missed opportunity at a great meal to death. We begin by assigning numeric values to the relative losses and storing these in a matrix with zeros along its diagonal:

```
➤ M = c(0,1, 2,100,0, 50,1,2,0)
➤ dim(M) = c(3,3
➤ rpart(species~CapLth + Volvath + Capwth + Volvasiz,
  parms=list(loss=M)))
  1) root 150 150 A.verna (0.3333333 0.3333333 0.3333333)
    2) Volvasiz< 0.8 50  0 A.citrina (1.0000000
         0.0000000 0.0000000) *
    3) Volvasiz>=0.8 100 100 A.verna (0.0000000
         0.5000000 0.5000000)
      6) Volvasiz< 1.85 66 32 A.verna (0.0000000
           0.7575758 0.2424242)
        12) Volvath< 5.3 58 16 A.verna (0.0000000
             0.8620690 0.1379310) *
        13) Volvath>=5.3 8  0 A.calyptroderma (0.0000000
             0.0000000 1.0000000) *
      7) Volvasiz>=1.85 34  0 A.calyptroderma
           (0.0000000 0.0000000 1.0000000) *
```

Note that the only assignments to edible species (the group of three values in parentheses at the end of each line) made in this decision tree are those which were unambiguous in the original sample. All instances, which by virtue of their measurements, might be confused with the extremely poisonous *Amanita verna* are classified as *Amanita verna*. A wise decision indeed!

Exercise 9.4: Develop a decision tree for classifying breast tumors as benign or malignant using the data at http://archive.ics.uci.edu/ml/datasets/Breast+Cancer+Wisconsin+(Diagnostic).

Account for the relative losses of misdiagnosis.

9.3 REGRESSION

Regression combines two ideas with which we gained familiarity in previous chapters:

1. Correlation or dependence among variables
2. Additive model.

Here is an example: Anyone familiar with the restaurant business (or indeed, with any number of businesses which provide direct service to the public, including the Post Office) knows that the volume of business is a function of the day of the week. Using an *additive model*, we can represent business volume via the formula

$$V_{ij} = \mu + \delta_i + z_{ij},$$

where V_{ij} is the volume of business on the ith day of the jth week, μ is the average volume, δ_i is the deviation from the average volume observed on the ith day of the week, $i = 1, \ldots, 7$, and the z_{ij} are independent, identically distributed random fluctuations.

Many physiological processes such as body temperature have a circadian rhythm, rising and falling each 24 hours. We could represent temperature by the formula

$$T_{ij} = \mu + \delta_i + z_{ij},$$

where i (in minutes) takes values from 1 to 24*60, but this would force us to keep track of 1441 different parameters. Besides, we can get almost as good a fit to the data using a formula with just two parameters, μ and β as in Equation 9.1:

$$E(T_{ij}) = \mu + \beta \cos(2\pi*(t + 300)/1440). \tag{9.1}$$

If you are not familiar with the cos() trigonometric function, you can use R to gain familiarity as follows:

```
➤ hour = c(0:24)
➤ P = cos(2*pi*(hour + 6)/24)
➤ plot(P)
```

Note how the cos() function first falls then rises, undergoing a complete cycle in a 24-hour period.

Why use a formula as complicated as Equation 9.1? Because now we have only two parameters we need to estimate, μ and β. For predicting body temperature $\mu = 98.6$ and $\beta = 0.4$ might be reasonable choices, though, of course, the values of these parameters will vary from individual to individual. For the author of this text, his mean body temperature is $\mu = 97.6°F$.

Exercise 9.5: If $E(Y) = 3X + 2$, can X and Y be independent?

Exercise 9.6: According to the inside of the cap on a bottle of Snaple's Mango Madness, "the number of times a cricket chirps in 15 seconds plus 37 will give you the current air temperature." How many times would you expect to hear a cricket chirp in 15 seconds when the temperature is 39°? 124°?

Exercise 9.7: If we constantly observe large values of one variable, call it Y, whenever we observe large values of another variable, call it X, does this mean X is part of the mechanism responsible for increases in the value of Y? If not, what are the other possibilities? To illustrate the several possibilities, give at least three real-world examples in which this statement would be false. (You'll do better at this exercise if you work on it with one or two others.)

9.3.1 Linear Regression

Equation 9.1 is an example of linear regression. The general form of linear regression is

$$Y = \mu + \beta f[X] + Z, \tag{9.2}$$

where Y is known as the *dependent* or *response variable*, X is known as the *independent variable* or *predictor*, $f[X]$ is a function of known form, μ and β are unknown *parameters*, and Z is a random variable whose expected value is zero. If it weren't for this last random component Z, then if we knew the parameters μ and β, we could plot the values of the dependent variable Y and the function $f[X]$ as a straight line on a graph. Hence the name, *linear regression*.

For the past year, the price of homes in my neighbor could be represented as a straight line on a graph relating house prices to time. $P = \mu + \beta t$, where μ was the price of the house on the first of the year and t is the day of the year. Of course, as far as the price of any individual house was concerned, there was a lot of fluctuation around this line depending on how good a salesman the realtor was and how desperate the owner was to sell.

If the price of my house ever reaches \$700 K, I might just sell and move to Australia. Of course, a straight line might not be realistic. Prices have a way of coming down as well as going up. A better prediction formula might be $P = \mu + \beta t - \gamma t^2$, in which prices continue to rise until $\beta - \gamma t = 0$, after which they start to drop. If I knew what β and γ were or could at least get some good estimates of their value, then I could sell my house at the top of the market!

The trick is to look at a graph such as Figure 9.1 and somehow extract that information.

Note that $P = \mu + \beta t - \gamma t^2$ is another example of *linear regression*, only with three parameters rather than two. So is the formula $W = \mu + \beta H + \gamma A + Z$, where W denotes the weight of a child, H is its height, A its age, and Z, as always, is a purely random component. $W = \mu + \beta H + \gamma A + \delta AH + Z$ is still another example. The parameters μ, β, γ, and so forth are sometimes referred to as the *coefficients* of the model.

What then is a *nonlinear* regression? Here are two examples:

$$Y = \beta \log(\gamma X),$$

which is linear in β but nonlinear in γ, and

$$T = \beta \cos(t + \gamma),$$

which also is linear in β but nonlinear in γ.

Regression models that are nonlinear in their parameters are beyond the scope of this text. The important lesson to be learned from their existence is that we need to have some idea of the functional relationship between a response variable and its predictors before we start to fit a linear regression model.

Exercise 9.8: Use R to generate a plot of the function $P = 100 + 10t - 1.5t^2$ for values of $t = 0, 1, \ldots 10$. Does the curve reach a maximum and then turn over?

9.4 FITTING A REGRESSION EQUATION

Suppose we have determined that the *response variable Y* whose value we wish to predict is related to the value of a *predictor variable X*, in that its expected value EY is equal to $a + bX$. On the basis of a sample of n pairs of observations (x_1, y_1), $(x_2, y_2), \ldots (x_n, y_n)$ we wish to estimate the unknown coefficients a and b. Three methods of estimation are in common use: ordinary least squares, error-in-variable or Deming regression, and least absolute deviation. We will study all three in the next few sections.

9.4.1 Ordinary Least Squares

The ordinary least squares (OLS) technique of estimation is the most commonly used, primarily for historical reasons, as its computations can be done (with some effort) by hand or with a primitive calculator. The objective of the method is to determine the parameter values that will minimize the sum of squares $\Sigma(y_i - EY)^2$, where the expected value of the dependent variable, EY, is modeled by the right-hand side of our regression equation.

In our example, $EY = a + bx_i$, and so we want to find the values of a and b that will minimize $\Sigma(y_i - a - bx_i)^2$. We can readily obtain the desired estimates with the aid of the R function **lm()** as in the following program, in which we regress systolic blood pressure (SBP) as a function of age (Age):

```
➢ Age = c(39,47,45,47,65,46,67,42,67,56,64,56,59,34,42)
➢ SBP = c(144,220,138,145,162,142,170,124,158,154,162,1
  50,140,110,128)
➢ lm(SBP~Age)
```

Coefficients:

```
(Intercept)        Age
     95.613      1.047
```

which we interpret as

$$SBP = \hat{a} + \hat{b}\text{Age} = 95.6 + 1.05\text{Age}.$$

Notice that when we report our results, we drop decimal places that convey a false impression of precision.

How good a fit is this regression line to our data? The R function resid() provides a printout of the *residuals*, that is, of the differences between the values our equation would predict and what individual values of SBP were actually observed.

```
➢ resid(lm(SBP~Age))
$residuals
   [1]    7.5373843  75.1578758  -4.7472470   0.1578758
  -1.6960182  -1.7946856
   [7]    4.2091047 -15.6049314  -7.7908953  -0.2690712
  -0.6485796  -4.2690712
  [13]  -17.4113868 -21.2254229 -11.6049314
```

Consider the fourth residual in the series, 0.15. This is the difference between what was observed, SBP = 145, and what the regression line estimates as the expected SBP for a 47-year-old individual E(SBP) = 95.6 + 1.04*47 = 144.8. Note that we were furthest off the mark in our predictions, that is, we had the largest residual, in the case of the youngest member of our sample, a 34-year old. Oops, I made a mistake; there is a still larger residual.

Mistakes are easy to make when glancing over a series of numbers with many decimal places. You won't make a similar mistake if you first graph your results using the following R code:

```
➢ Pred = lm(SBP ~ Age)
➢ plot (Age, SBP)
➢ lines (Age,fitted(Pred)).
```

Now the systolic blood pressure value of 220 for the 47-year-old stands out from the rest.

The linear model R function **lm()** lends itself to the analysis of more complex regression functions. In this example and the one following, we use functions as predictors. For example, if we had observations on body temperature taken at various times of the day, we would write*

```
➢ lm (temp ~ I(cos (2*π*(time + 300)/1440)))
```

when we wanted to obtain a regression formula. Or, if we had data on weight and height as well as age and systolic blood pressure, we would write

```
➢ lm(SBP ~ Age+I(100*(weight/(height*height)))).
```

If our regression model is $EY = \mu + \alpha X + \beta W$, that is, if we have two *predictor variables* X and W, we would write $lm(Y \sim X + W)$. If our regression model is

---

* To ensure that the arithmetic operators have their accustomed interpretations—rather than those associated with the lm() function—it is necessary to show those functions as arguments to the **I( )** function.

$$EY = \mu + \alpha f[X] + \beta g[W],$$

we would write

$$lm(Y \sim I(f[X]) + I(g[W])).$$

Note that the lm() function automatically provides an estimate of the *intercept* μ. The term intercept stands for "intercept with the Y-axis," since when $f[X]$ and $g[W]$ are both zero, the expected value of Y will be μ.

Sometimes, we know in advance that μ will be zero. Two good examples would be when we measure distance traveled or the amount of a substance produced in a chemical reaction as functions of time. We can force lm() to set $\mu = 0$ by writing lm($Y \sim 0 + X$).

Sometimes, we would like to introduce a cross-product or interaction term into the model, $EY = \mu + \alpha X + \beta W + \gamma XW$, in which case, we would write lm($Y \sim X*W$). Note that when used in expressing a linear model formula, the arithmetic operators in R have interpretations different to their usual ones. Thus, in $Y \sim X*W$, the "*" operator defines the regression model as $EY = \mu + \alpha X + \beta W + \gamma XW$ and not as $EY = \mu + \gamma XW$ as you might expect. If we wanted the latter model, we would have to write $Y \sim I(X*W)$. The I( ) function restores the usual meanings to the arithmetic operators.

Exercise 9.9: Do U.S. residents do their best to spend what they earn? Fit a regression line using the OLS to the data in the accompanying table relating disposable income to expenditures in the United States from 1960 to 1982.

Economic Report of the President, 1988, Table B-27		
	Income	Expenditures
Year	1982 $s	1982 $s
1960	6036	5561
1962	6271	5729
1964	6727	6099
1966	7280	6607
1968	7728	7003
1970	8134	7275
1972	8562	7726
1974	8867	7826
1976	9175	8272
1978	9735	8808
1980	9722	8783
1982	9725	8818

Exercise 9.10: Suppose we've measured the dry weights of chicken embryos at various intervals at gestation and recorded our findings in the following table:

Age in days	6	7	8	9	10	11	12	13	14	15	16
Weight (gms)	0.029	0.052	0.079	0.125	0.181	0.261	0.425	0.738	1.130	1.882	2.812

Create a scatter plot for the data in the table and overlay this with a plot of the regression line of weight with respect to age. Recall from Section 6.1 that the preferable way to analyze growth data is by using the logarithms of the exponentially increasing values. Overlay your scatter plot with a plot of the new regression line of log(weight) as a function of age. Which line (or model) appears to provide the better fit to the data?

Exercise 9.11: Obtain and plot the OLS regression of systolic blood pressure with respect to age after discarding the outlying value of 220 recorded for a 47-year-old individual.

Exercise 9.12: In a further study of systolic blood pressure as a function of age, the height and weight of each individual were recorded. The latter were converted to a Quetlet index using the formula $QUI = 100*weight/height^2$. Fit a multivariate regression line of systolic blood pressure with respect to age and the Quetlet index using the following information:

Age = c(41, 43, 45, 48, 49, 52, 54, 56, 57, 59, 62, 63, 65)
SBP = c(122, 120, 135, 132, 130, 148, 146, 138, 135, 166, 152, 170, 164)
QUI = c(3.25, 2.79, 2.88, 3.02, 3.10, 3.77, 2.98, 3.67, 3.17, 3.88, 3.96, 4.13, 4.01)

9.4.2 Types of Data

The linear regression model is a quantitative one. When we write $Y = 3 + 2X$, we imply that the product $2X$ will be meaningful. This will be the case if X is a continuous measurement like height or weight. In many surveys, respondents use a nine-point *Likert scale*, where a value of "1" means they definitely disagree with a statement, and "9" means they definitely agree. Although such data are ordinal and not continuous or metric, the regression equation is still meaningful.

When one or more predictor variables are categorical, we must use a different approach. R makes it easy to accommodate categorical predictors. We first declare the categorical variable to be a factor as in Section 1.6. The regression model fit by the **lm()** function will include a different additive component for each level of the factor or categorical variable. Thus, we can include sex or race as predictors in a regression model.

Exercise 9.13: Using the Milazzo data of Exercise 6.20, express the fecal coliform level as a linear function of oxygen level and month. Solve for the coefficients using OLS.

9.4.3 Least Absolute Deviation Regression

Least absolute deviation regression (LAD) attempts to correct one of the major flaws of OLS, that of giving sometimes excessive weight to extreme values. The LAD method solves for those values of the coefficients in the regression equation for which the sum of the absolute deviations $\Sigma |y_i - R[x_i]|$ is a minimum. You'll need to install a new package in R in order to do the calculations.

The easiest way to do an installation is get connected to the Internet and then type

> ➤ install.packages("quantreg")

The installation which includes downloading, unzipping, and integrating the new routines is then done automatically. The installation needs to be done once and once only. But each time before you can use the LAD routine, you'll need to load the supporting functions into computer memory by typing

> ➤ library(quantreg)

To perform the LAD regression, type

> ➤ rq(formula = SBP ~ Age)

to obtain the following output:
    Coefficients:

```
(Intercept)        Age
  81.157895    1.263158
```

To display a graph, type

```
➤ f = coef(rq(SBP ~ Age))
➤ pred = f[1] + f[2]*Age
➤ plot (Age, SBP)
➤ lines (Age, pred)
```

Exercise 9.14: Create a single graph on which both the OLS and LAD regressions for the Age and SBP are displayed.

Exercise 9.15: Repeat the previous exercise after first removing the outlying value of 220 recorded for a 47–ld individual.

Exercise 9.16: Use the LAD method using the data of Exercise 9.7 to obtain a prediction equation for systolic blood pressure as a function of age and the Quetlet index.

Exercise 9.17: The State of Washington uses an audit recovery formula in which the percentage to be recovered (the overpayment) is expressed as a linear function of the amount of the claim. The slope of this line is close to zero. Sometimes, depending on the audit sample, the slope is positive, and sometimes it is negative. Can you explain why?

9.4.4 Errors-in-Variables Regression

The need for errors-in-variables (EIV) or Deming regression is best illustrated by the struggles of a small medical device firm to bring its product to market. Their first challenge was to convince regulators that their long-lasting device provided equivalent results to those of a less-efficient device already on the market. In other words, they needed to show that the values V recorded by their device bore a linear relation to the values W recorded by their competitor, that is, that $V = a + bW + Z$, where Z is a random variable with mean 0.

In contrast to the examples of regression we looked at earlier, the errors inherent in measuring W (the so-called predictor) were as large if not larger than the variation inherent in the output V of the new device.

The EIV regression method they used to demonstrate equivalence differs in two respects from that of OLS:

1. With OLS, we are trying to minimize the sum of squares $\Sigma(y_{oi} - y_{pi})^2$, where y_{oi} is the ith observed value of Y, and y_{pi} is the ith predicted value. With EIV, we are trying to minimize the sums of squares of errors, going both ways: $\Sigma(y_{oi} - y_{pi})^2/\mathrm{Var}Y + \Sigma(x_{oi} - x_{pi})^2/\mathrm{Var}X$.

2. The coefficients of the EIV regression line depend on $\lambda = \mathrm{Var}X/\mathrm{Var}Y$.

To compute the EIV regression coefficients, you'll need the following R function along with an estimate of the ratio lambda of the variances of the two sets of measurements:

```
➤ eiv = function (X,Y, lambda){
➤ My = mean(Y)
➤ Mx = mean (X)
➤ Sxx = var(X)
➤ Syy = var(Y)
➤ Sxy = cov(X,Y)
➤ diff = lambda*Syy-Sxx
➤ root = sqrt(diff*diff+4*lambda*Sxy*Sxy)
➤ b = (diff+ root)/(2*lambda*Sxy)
➤ a = My - b*Mx
➤ list (a=a, b=b)
➤ }
```

Exercise 9.18: Are these two sets of measurements comparable, that is, does the slope of the EIV regression line differ significantly from unity? Hint: Use the sample variances to estimate lambda.

$X = c(2.521, 3.341, 4.388, 5.252, 6.422, 7.443, 8.285, 9.253, 10.621,$
$\quad 10.405, 11.874, 13.444, 13.343, 16.402, 19.108, 19.25, 20.917,$
$\quad 23.409, 5.583, 5.063, 6.272, 7.469, 10.176, 6.581, 7.63)$

$Y = c(2.362, 3.548, 4.528, 4.923, 6.443, 6.494, 8.275, 9.623, 9.646,$
$\quad 11.542, 10.251, 11.866, 13.388, 17.666, 17.379, 21.089, 21.296,$
$\quad 23.983, 5.42, 6.369, 7.899, 8.619, 11.247, 7.526, 7.653)$

Exercise 9.19: Which method should be used to regress U as a function of W in the following cases, OLS, LAD, or EIV?

a. Some of the U values are suspect.

b. It's not clear whether U or W is the true independent variable or whether both depend on the value of a third hidden variable.

c. Minor errors in your predictions aren't important; large ones could be serious.

9.4.5 Assumptions

To use any of the preceding linear regression methods, the following as-yet-unstated assumptions must be satisfied:

1. **Independent Random Components.** In the model $y_i = \mu + \beta x_i + z_i$, the random fluctuations z_i must be independent of one another. If the z_i are not, it may be that a third variable W is influencing their values. In such a case, we would be advised to try the model $y_i = \mu + \beta x_i + \gamma w_i + \varepsilon_i$.

   When observations are made one after the other in time, it often is the case that successive errors are dependent on one another, but we can remove this dependence if we work with the increments $y2 - y1$, $y3 - y2$, and so forth. Before we start fitting our model, we would convert to these differences using the following R code:

   ```
   > xdiff = c(1:length(x) -1)
   > ydiff = c(1:length(x) -1)
   > for (i in 1:length(x)-1){
   > + xdiff[i]= x[i+1] - x[i]
   > + ydiff[i]= y[i+1] - y[i]
   > }
   ```

2. **Identically Distributed Random Components.** Often, it is the case that large random fluctuations are associated with larger values of the observations. In such cases, techniques available in more advanced textbooks provide for weighting the observations when estimating model coefficients so that the least variable observations receive the highest weight.

3. **Random Fluctuations Follow a Specific Distribution.** Most statistics packages provide tests of the hypotheses that the model parameters are significantly different from zero. For example, if we enter the following instructions in R

   ```
   > Age = c(39,47,45,47,65,46,67,42,67,56,64,56,59,34,
   42)
   > SBP = c(144,220,138,145,162,142,170,124,158,154,16
   2,150,140,110,128)
   > summary(lm(SBP~Age))
   ```

A lengthy output is displayed, which includes the statements

Coefficients:

	Estimate	Std. Error	t value	Pr(>|t|)	
(Intercept)	95.6125	29.8936	3.198	0.00699	**
Age	1.0474	0.5662	1.850	0.08717	

This tells us that the intercept of the OLS regression is significantly different from 0 at the 1% level, but the coefficient for Age is not significant at the 5% level. The p-value 0.08717 is based on the assumption that all the random fluctuations come from a normal distribution. If they do not, then this p-value may be quite misleading as to whether or not the regression coefficient of systolic blood pressure with respect to age is significantly different from 0.

An alternate approach is to obtain a confidence interval for the coefficient by taking a series of bootstrap samples from the collection of pairs of observations (39,144), (47,220), . . . , (42,128).

> **Exercise 9.20:** Obtain a confidence interval for the regression coefficient of systolic blood pressure with respect to age using the bias-corrected-and-accelerated bootstrap described in Section 4.2.4.

9.5 PROBLEMS WITH REGRESSION

At first glance, regression seems to be a panacea for all modeling concerns. But it has a number of major limitations, just a few of which we will discuss in this section.

- The model that best fits the data we have in hand may not provide the best fit to the data we gather in the future.
- More than one linear regression model may provide a statistically significant fit to our data.

9.5.1   Goodness of Fit versus Prediction

Whenever we use a regression equation to make predictions, we make two implicit assumptions:

1. Relationships among the variables remain unchanged over time.
2. The sources of variation are the same as when we first estimated the coefficients.

We are seldom on safe grounds when we attempt to use our model outside the range of predictor values for which it was developed originally. For one thing, literally every phenomenon seems to have nonlinear behavior for very small and very large values of the predictors. Treat every predicted value outside the original data range as suspect.

In my lifetime, regression curves failed to predict a drop in the sales of '78 records as a result of increasing sales of 45s, a drop in the sales of 45s as a result of increasing sales of 8-track tapes, a drop in the sales of 8-track tapes as a result of increasing sales of cassettes, nor the drop in the sales of cassettes as a result of increasing sales of CDs. It is always advisable to revalidate any model that has been in use for an extended period (see Section 9.5.2).

Finally, let us not forgot that our models are based on samples, and that sampling error must always be inherent in the modeling process.

Exercise 9.21: Redo Exercise 9.7.

9.5.2   Which Model?

The exact nature of the formula connecting two variables cannot be determined by statistical methods alone. If a linear relationship exists between two variables X and Y, then a linear relationship also exists between Y and any monotone (nondecreasing or nonincreasing) function of X. Assume X can only take positive values. If we can fit Model I: $Y = \alpha + \beta X + \varepsilon$ to the data, we also can fit Model II: $Y = \alpha' + \beta'\log[X] + \varepsilon$, and Model III: $Y = \alpha'' + \beta''X + \gamma X^2 + \varepsilon$. It can be very difficult to determine which model if any is the "correct" one.

Five principles should guide you in choosing among models:

1. Prevention. The data you collect should span the entire range of interest. For example, when employing EIV regression to compare two methods of measuring glucose, it is essential to observe many pairs of observed abnormal values (characteristic of a disease process) along with the more-readily available pairs of normal values.

2. Don't allow your model to be influenced by one or two extreme values—whenever this is a possibility, employ LAD regression rather than OLS. Strive to obtain response observations at intervals throughout the relevant range of the predictor. Only when we have observations spanning the range of interest can we begin to evaluate competitive models.

3. Think Why Rather Than What.   In Exercise 9.6, we let our knowledge of the underlying growth process dictate the use of $\log(X)$ rather than X. As a second example, consider that had we wanted to find a relationship between the volume V and temperature T of a gas, any of the preceding three models might have been used to fit the relationship. But only one, the model $V = a + KT$, is consistent with kinetic molecular theory.

4. Plot the *residuals,* that is, plot the error or difference between the values predicted by the model and the values that were actually observed. If a pattern emerges from the plot, then modify the model to correct for the pattern. The next two exercises illustrate this approach.

Exercise 9.22: Apply Model III to the blood pressure data at the beginning of Section 9.3.1. Use R to plot the residuals with the code:

```
➤ temp=lm(SBP~Age+Age*Age)
➤ plot(Age,resid(temp))
```

What does this plot suggest about the use of Model III in this context versus the simpler model that was used originally?

Exercise 9.23: Plot the residuals for the models and data of Exercise 7.6. What do you observe?

The final two guidelines are contradictory in nature:

5. *The More Parameters, the Better the Fit.* Thus, Model III is to be preferred to the two simpler models.

6. *The simpler, more straightforward model* is more likely to be correct when we come to apply it to data other than the observations in hand; thus, Models I and II are to be preferred to Model III.

9.5.3 Measures of Predictive Success

In order to select among models, we need to have some estimate of how successful each model will be for prediction purposes. One measure of the goodness-of-fit of the model is

$$\text{SSE} = \sum (y_i - y_i^*)^2,$$

where y_i and y_i^* denote the ith observed value and the corresponding value obtained from the model. The smaller this sum of squares, the better the fit.

If the observations are independent, then

$$\sum (y_i - y_i^*)^2 = \sum (y_i - \bar{y})^2 - \sum (\bar{y} - y_i^*)^2.$$

The first sum on the right hand side of the equation is the total sum of squares (SST). Most statistics software uses as a measure of fit $R^2 = 1 - \text{SSE/SST}$. The closer the value of R^2 is to 1, the better.

The automated entry of predictors into the regression equation using R^2 runs the risk of overfitting, as R^2 is guaranteed to increase with each predictor entering the model. To compensate, one may use the adjusted R^2

$$1 - [((n - i)(1 - R^2))/(n - p)],$$

where n is the number of observations used in fitting the model, p is the number of regression coefficients in the model, and i is an indicator variable that is 1 if the model includes an intercept, and 0 otherwise.

When you use the R function **summarize(lm())**, it calculates values of both R-squared (R^2) and adjusted R^2.

9.5.4 Multivariable Regression

We've already studied several examples in which we utilized multiple predictors in order to obtain improved models. Sometimes, as noted in Section 9.4.5, dependence among the random errors (as seen from a plot of the residuals) may force us to use additional variables. This result is in line with our discussion of experimental design in Chapter 6. Either we must control each potential source of variation, measure it, or tolerate the "noise."

But adding more variables to the model equation creates its own set of problems. Do predictors U and V influence Y in a strictly additive fashion so that we may write $Y = \mu + \alpha U + \beta V + z$ or, equivalently, $\text{lm}(Y \sim U + V)$. What if U represents the amount of fertilizer, V the total hours of sunlight, and Y is the crop yield? If there are too many cloudy days, then adding fertilizer won't help a bit. The effects of fertilizer and sunlight are superadditive (or *synergistic*). A better model would be $Y = \mu + \alpha U + \beta V + \gamma UV + z$ or $\text{lm}(Y \sim U*V)$.

To achieve predictive success, our observations should span the range over which we wish to make predictions. With only a single predictor, we might make just 10 observations spread across the predictor's range. With two synergistic or antagonist predictors, we are forced to make 10×10 observations, with three, $10 \times 10 \times 10 = 1000$ observations, and so forth. We can cheat, scattering our observations at random or in some optimal systematic fashion across the grid of possibilities, but there will always be a doubt as to our model's behavior in the unexplored areas.

The vast majority of predictors are interdependent. Changes in the value of 1 will be accompanied by changes in the other. (Notice that we do *not* write, "Changes in the value of one will cause or result in changes in the other." There may be yet a third, hidden variable responsible for all the changes.) What this means is that more than one set of coefficients may provide an equally good fit to our data. And more than one set of predictors!

As the following exercise illustrates, whether or not a given predictor will be found to make a statistically significant contribution to a model will depend upon what other predictors are present.

Exercise 9.24: Utilize the data from the Consumer Survey to construct a series of models as follows:

Express Purchase as a function of Fashion and Gamble

Express Purchase as a function of Fashion, Gamble, and Ozone

Express Purchase as a function of Fashion, Gamble, Ozone, and Pollution

In each instance, use the summary function along with the linear model function, as in **summary(lm())**, to determine the values of the coefficients, the associated p-values, and the values of multiple R^2 and adjusted R^2. (As noted in the preface, for your convenience, the following datasets may be downloaded from http://statcourse/Data/ConsumerSurvey.htm.)

As noted earlier, when you use the R function **summary(lm())**, it yields values of multiple R^2 and adjusted R^2. Both are measures of how satisfactory a fit the model provides to the data. The second measure adjusts for the number of terms in the model; as this exercise reveals, the greater the number of terms, the greater the difference between the two measures.

The R function **step()** automates the variable selection procedure. Unless you specify an alternate method, **step()** will start the process by including all the variables you've listed in the model. Then it will eliminate the least significant predictor, continuing a step at a time until all the terms that remain in the model are significant at the 5% level.

Applied to the data of Exercise 9.24, one obtains the following results:

```
➢ summary(step(lm(Purchase~Fashion + Gamble + Ozone
  + Pollution)))
Start: AIC = 561.29
 Purchase  ~  Fashion + Gamble + Ozone + Pollution
            Df   Sum of Sq RSS      AIC
- Ozone      1    1.82       1588.97 559.75
- Pollution 1    2.16       1589.31 559.84
<none>                       1589.15 561.29
- Fashion    1    151.99     1739.14 595.87
- Gamble     1    689.60     2276.75 703.62
Step: AIC = 559.75
 Purchase ~ Fashion + Gamble + Pollution
            Df   Sum of Sq RSS      AIC
<none>                       1588.97 559.75
- Pollution 1    17.08      1606.05  562.03
- Fashion    1    154.63     1743.60  594.90
- Gamble     1    706.18     2295.15  704.84
```

Call:

```
lm(formula = Purchase ~ Fashion + Gamble + Pollution)
Residuals:
  Min       1Q        Median    3Q       Max
 -5.5161  -1.3955  -0.1115    1.4397   5.0325
```

Coefficients:

```
               Estimate Std. Error  t value  Pr(>|t|)
(Intercept) -1.54399  0.49825      -3.099   0.00208 **
Fashion      0.41960  0.06759       6.208   1.36e-09 ***
Gamble       0.85291  0.06429      13.266   <2e-16 ***
Pollution    0.14047  0.06809       2.063   0.03977 *
―
Signif. codes: 0 `***' 0.001 `**' 0.01 `*' 0.05 `.' 0.1
  ` ' 1
Residual standard error: 2.003 on 396 degrees of
freedom
Multiple R-Squared: 0.4051, Adjusted R-squared: 0.4005
F-statistic: 89.87 on 3 and 396 DF, p-value: <2.2e-16
```

From which we extract the model

```
E(Purchase) = -1.54 + 0.42*Fashion + 0.85*Gamble +
0.14*Pollution.
```

In terms of the original questionnaire, this means that the individuals most likely to purchase an automobile of this particular make and model are those who definitely agreed with the statements "Life is too short not to take some gambles," "When I must choose, I dress for fashion, not comfort," and "I think the government is doing too much to control pollution."

> **Exercise 9.25:** The marketing study also included 11 more questions, each using a nine-point Likert scale. Use the R stepwise regression function to find the best prediction model for Attitude using the answers to all 15 questions.

9.6 QUANTILE REGRESSION

Linear regression techniques are designed to help us predict expected values, as in $EY = \mu nn + \beta X$. But what if our real interest is in predicting extreme values, if, for example, we would like to characterize the observations of Y that are likely to lie in the upper and lower tails of Y's distribution.

Even when expected values or medians lie along a straight line, other quantiles may follow a curved path. Koenker and Hallock applied the method of quantile regression to data taken from Ernst Engel's study in 1857 of the dependence of households' food expenditure on household income. As Figure 9.7 reveals, not only was an increase in food expenditures observed as expected when household income was increased, but the dispersion of the expenditures increased also.

In estimating the τth quantile,* we try to find that value of β, for which

$$\Sigma_k \rho_\tau (y_k - f[x_k, \beta])$$

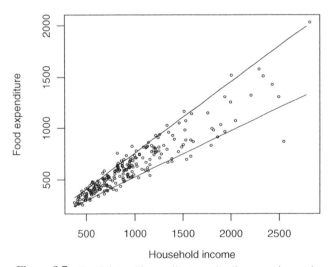

Figure 9.7 Engel data with quantile regression lines superimposed.

---

* τ is pronounced tau.

is a minimum, where

$$\rho_\tau[x] = \tau x \qquad \text{if } x > 0$$
$$= (\tau - 1)x \quad \text{if } x \leq 0.$$

For the 50th percentile or median, we have LAD regression. In fact, we make use of the **rq**() function to obtain the LAD regression coefficients.

Begin by loading the Engel data, which are included in the quantreg library, and plotting the data.

```
➤ library(quantreg)
➤ data(engel)
➤ attach(engel)
➤ plot(x,y,xlab="household income",ylab="food
  expenditure",cex=.5)
```

One of the data points lies at quite a distance from the others. Let us remove it for fear it will have too strong an influence on the resulting regression lines.

```
➤ x1 = engel$x[-138]
➤ y1 = engel$y[-138]
```

Finally, we overlay the plot with the regression lines for the 10th and 90th percentiles:

```
➤ plot(x1,y1,xlab="household income",ylab="food
  expenditure",cex=.5)
➤ taus = c(.1,.9)
➤ c = (max(x1)-min(x1))/100
➤ xx = seq(min(x1),max(x1),by=c)
➤ for(tau in taus){
+    f = coef(rq((y1)~(x1), tau=tau))
+    yy = (f[1] + f[2]*(xx))
+    lines(xx,yy)
+    }
```

Exercise 9.26: Plot the 10th and 90th quantile regression lines for the Engel economic data. How did removing that one data point affect the coefficients?

Exercise 9.27: Brian Cady has collected data relating acorn yield (WTPERHA) to the condition of the oak canopy (OAKCCSI). Plot the 25th, 50th, and 75th quantile regression lines for the following data:

WTPERHA = c(43.04055, 4.946344, 60.21557, 81.66009, 50.61638, 56.65638,
60.92875, 72.11634, 72.80992, 32.66437, 202.776, 116.7866,
14.56848, 59.79613, 37.25652, 93.01928, 20.869, 46.28904,
50.56929, 19.39652, 71.11756, 16.27327, 11.51406, 67.86525,
80.7973, 25.6447, 108.2565, 49.49308, 76.17169, 75.50679,
32.31519, 36.78138, 37.04142, 49.99904, 12.82474, 73.1297,
20.20769, 1.11135, 8.10927, 50.85466, 89.92351, 16.45993, 29.18572)

OAKCCSI = c(0.826136, 0.115109, 0.447214, 0.9, 0.52915, 1, 0.708872, 0.864581,

       0.812404, 0.777817, 0.781025, 1, 0.380789, 0.54658, 0.482183,

       1, 0.455522, 0.83666, 0.648074, 0.860233, 0.61441, 0.316228,

       0.626498, 0.89861, 0.597913, 0.33541, 0.953939, 0.972111, 1, 1,

       0.927362, 0.924662, 0.642262, 0.667458, 0.47697, 0.91515,

       0.589491, 0.766485, 0.259808, 0.606218, 0.67082, 0.736546, 0.8)

9.7 VALIDATION

As noted in the preceding sections, more than one model can provide a satisfactory fit to a given set of observations; even then, goodness of fit is no guarantee of predictive success. Before putting the models we develop to practical use, we need to *validate* them. There are three main approaches to validation:

1. Independent verification (obtained by waiting until the future arrives or through the use of surrogate variables).

2. Splitting the sample (using one part for calibration or training, the other for verification or testing).

3. Resampling (taking repeated samples from the original sample and refitting the model each time).

In what follows, we examine each of these methods in turn.

9.7.1 Independent Verification

Independent verification is appropriate and preferable whatever the objectives of your model. In geologic and economic studies, researchers often return to the original setting and take samples from points that have been bypassed on the original round. In other studies, verification of the model's form and the choice of variables are obtained by attempting to fit the same model in a similar but distinct context.

For example, having successfully predicted an epidemic at one army base, one would then wish to see if a similar model might be applied at a second and third almost-but-not-quite identical base.

Independent verification can help discriminate among several models that appear to provide equally good fits to the data. Independent verification can be used in conjunction with either of the two other validation methods. For example, an automobile manufacturer was trying to forecast parts sales. After correcting for seasonal effects and long-term growth within each region, autoregressive integrated moving average (ARIMA) techniques were used.* A series of best-fitting ARIMA models was derived: one model for each of the nine sales regions into which the

---

* For examples and discussion of ARIMA processes used to analyze data whose values change with time, see Brockwell and Davis, *Time Series Theory and Methods* (Springer-Verlag, 1987).

sales territory had been divided. The nine models were quite different in nature. As the regional seasonal effects and long-term growth trends had been removed, a single ARIMA model applicable to all regions, albeit with coefficients that depended on the region, was more plausible. The model selected for this purpose was the one that gave the best fit on average when applied to all regions.

Independent verification also can be obtained through the use of surrogate or proxy variables. For example, we may want to investigate past climates and test a model of the evolution of a regional or worldwide climate over time. We cannot go back directly to a period before direct measurements on temperature and rainfall were made, but we can observe the width of growth rings in long-lived trees or measure the amount of carbon dioxide in ice cores.

9.7.2 Splitting the Sample

For validating time series, an obvious extension of the methods described in the preceding section is to hold back the most recent data points, fit the model to the balance of the data, and then attempt to "predict" the values held in reserve.

When time is not a factor, we still would want to split the sample into two parts, one for estimating the model parameters, the other for verification. The split should be made at random. The downside is that when we use only a portion of the sample, the resulting estimates are less precise.

In the following exercises, we want you to adopt a compromise proposed by C. I. Moiser.* Begin by splitting the original sample in half; choose your regression variables and coefficients independently for each of the subsamples. If the results are more or less in agreement, then combine the two samples and recalculate the coefficients with greater precision.

There are several different ways to program the division in R. Here is one way:

Suppose we have 100 triples of observations Y, P1, and P2. Using M and V to denote the values, we select for use in estimating and validating respectively,

```
➤ select = sample(100,50,replace=FALSE)
➤ YM = Y[select]
➤ P1M = P1[select]
➤ P2M = P2[select]
#exclude the points selected originally
➤ YV = Y[-select]
➤ P1V = P1[-select]
➤ P2V = P2[-select]
```

Exercise 9.28: Apply Moiser's method to the Milazzo data of the previous chapter. Can total coliform levels be predicted on the basis of month, oxygen level, and temperature?

---

* Symposium: The need and means of cross-validation. I: Problems and design of cross-validation. *Educat. Psych. Measure.* 1951; 11: 5–11.

Exercise 9.29: Apply Moiser's method to the data provided in Exercises 9.24 and 9.25 to obtain prediction equation(s) for Attitude in terms of some subset of the remaining variables.

Note: As conditions and relationships do change over time, any method of prediction should be *revalidated* frequently. For example, suppose we had used observations from January 2000 to January 2004 to construct our original model and held back more recent data from January to June 2004 to validate it. When we reach January 2005, we might refit the model, using the data from 1/2000 to 6/2004 to select the variables and determine the values of the coefficients, then use the data from 6/2004 to 1/2005 to validate the revised model.

Exercise 9.30: Some authorities would suggest discarding the earliest observations before refitting the model. In the present example, this would mean discarding all the data from the first half of the year 2000. Discuss the possible advantages and disadvantages of discarding these data.

9.7.3   Cross-Validation with the Bootstrap

Recall that the purpose of bootstrapping is to simulate the taking of repeated samples from the original population (and to save money and time by not having to repeat the entire sampling procedure from scratch). By bootstrapping, we are able to judge to a limited extent whether the models we derive will be useful for predictive purposes or whether they will fail to carry over from sample to sample. As the next exercise demonstrates, some variables may prove more reliable and consistent as predictors than others.

Exercise 9.31: Bootstrap repeatedly from the data provided in Exercises 9.24 and 9.25 and use the R **step**() function to select the variables to be incorporated in the model each time. Are some variables common to all the models?

9.8   SUMMARY AND REVIEW

In this chapter, you were introduced to two techniques for classifying and predicting outcomes: linear regression and classification and regression trees. Three methods for estimating linear regression coefficients were described along with guidelines for choosing among them. You were provided with a stepwise technique for choosing variables to be included in a regression model. The assumptions underlying the regression technique were discussed along with the resultant limitations. Overall guidelines for model development were provided.

You learned the importance of validating and revalidating your models before placing any reliance upon them. You were introduced to one of the simplest of pattern recognition methods, the classification and regression tree to be used whenever there are large numbers of potential predictors or when classification rather than quantitative prediction is your primary goal.

You were provided with R functions for the following:

- modify data values using cut(), seq()
- math including sqrt, max, min
- trig including cos() (yes, there are also a sin(), tan(), and cotan()
- statistics cov().

You discovered that certain R functions, print() is an example, yield different results depending on the type of object to which they are applied.

You learned how to download and install library packages not part of the original R download, including library(quantreg) and library(rpart). You made use of several modeling functions including lm and rq (with their associated functions fitted, summary, resid, step, and coef), and **rpart** (with its associated functions prune, predict, and print). You learned two new ways to fill an R vector and to delete selected elements. And you learned how to add lines and text to existing graphs.

Exercise 9.32: Make a list of all the italicized terms in this chapter. Provide a definition for each one along with an example.

Exercise 9.33: It is almost self-evident that levels of toluene, a commonly used solvent, would be observed in the blood after working in a room where the solvent was present in the air. Do the observations recorded below also suggest that blood levels are a function of age and bodyweight? Construct a model for predicting blood levels of toluene using these data.

Blood Tol	0.494	0.763	0.534	0.552	1.084	0.944	0.955	0.696
Air Tol	50	50	50	50	100	100	100	100
Weight	378	439	302	405	421	370	363	389
Age	95	95	84	85	86	86	83	86

Blood Tol	12.085	9.647	7.524	10.783	38.619	25.402	26.481	28.155
Air Tol	500	500	500	500	1000	1000	1000	1000
Weight	371	347	misg	348	378	433	363	420
Age	83	84	85	85	93	93	85	86

Exercise 9.34: Using the data from Exercise 6.18, develop a model for predicting whether an insurance agency will remain solvent.

Chapter 10

Reporting Your Findings

In this chapter, we assume you have just completed an analysis of your own or someone else's research and now wish to issue a report on the overall findings. You'll learn what to report and how to go about reporting it with particular emphasis on the statistical aspects of data collection and analysis.

> *One of the most common misrepresentations in scientific work is the scientific paper itself. It presents a mythical reconstruction of what actually happened. All mistaken ideas, false starts, badly designed experiments, and incorrect calculations are omitted. The paper presents the research as if it had been carefully thought out, planned, and executed according to a neat, rigorous process, for example, one involving testing of a hypothesis. The misrepresentation of the scientific paper is the most formal aspect of the misrepresentation of science as an orderly process based on a clearly defined method.*
>
> Brian Martin

10.1 WHAT TO REPORT

Reportable elements include all of the following:

- Study objectives
- Hypotheses
- Power and sample size calculations
- Data collection methods
- Validation methods
- Data summaries
- Details of the statistical analysis
- Sources of missing data
- Exceptions.

Introduction to Statistics Through Resampling Methods and R, Second Edition. Phillip I. Good.
© 2013 John Wiley & Sons, Inc. Published 2013 by John Wiley & Sons, Inc.

10.1.1 Study Objectives

If you are contributing to the design or analysis of someone else's research efforts, a restatement of the objectives is an essential first step. This ensures that you and the principal investigator are on the same page. This may be necessary in order to formulate quantifiable, testable hypotheses.

Objectives may have shifted or been expanded upon. Often such changes are not documented. You cannot choose or justify the choice of statistical procedures without a thorough understanding of study objectives.

10.1.2 Hypotheses

To summarize what was stated in Chapter 5, both your primary and alternate hypotheses must be put in quantified testable form. Your primary hypothesis is used to establish the significance level and your alternative hypothesis to calculate the power of your tests.

Your objective may be to determine whether adding a certain feature to a product would increase sales. (Quick. Will this alternative hypothesis lead to a one-sided or a two-sided test?) Yet, for reasons that have to do solely with the limitations of statistical procedures, your primary hypothesis will normally be a null hypothesis of no effect.

Not incidentally, as we saw in Chapter 8, the optimal statistical test for an ordered response is quite different from the statistic one uses for detecting an arbitrary difference among approaches. All the more reason why we need to state our alternative hypotheses explicitly.

10.1.3 Power and Sample Size Calculations

Your readers will want to know the details of your power and sample size calculations early on. If you don't let them know, they may assume the worst, for example, that your sample size is too small and your survey is not capable of detecting significant effects. State the alternative hypotheses that your study is intended to detect. Reference your methodology and/or the software you employed in making your power calculations. State the source(s) you relied on for your initial estimates of incidence and variability.

Here is one example. "Over a 10-year period in the Himalayas, Dempsy and Peters (1995) observed an incidence of five infected individuals per 100 persons per year. To ensure a probability of at least 90% of detecting a reduction in disease incidence from five persons to one person per 100 persons per year using a one-sided Fisher's exact test at the 2.5% significance level, 400 individuals were assigned to each experimental procedure group. This sample size was determined using the StatXact-5 power calculations for comparing two binomials."

10.1.4   Data Collection Methods

Although others may have established the methods of data collection, a comprehensive knowledge of these methods is essential to your choice of statistical procedures and should be made apparent in report preparation. Consider that 90% of all errors occur during data collection, as observations are erroneously recorded (garbage in, garbage out, GIGO), guessed at, or even faked. Seemingly innocuous workarounds may have jeopardized the integrity of the study. You need to know and report on exactly how the data were collected, not on how they were supposed to have been collected.

You need to record how study subjects were selected, what was done to them (if appropriate), and when and how this was done. Details of recruitment or selection are essential if you are to convince readers that your work is applicable to a specific population. If incentives were used (phone cards, t-shirts, cash), their use should be noted.

Readers will want to know the nature and extent of any blinding (and of the problems you may have had to overcome to achieve it). They will want to know how each subject was selected—random, stratified, or cluster sampling? They will want to know the nature of the controls (and the reasoning underlying your choice of a passive or active control procedure) and of the experimental design. Did each subject act as his own control as in a crossover design? Were case controls employed? If they were matched, how were they matched?

You will need the answers to all of the following questions and should incorporate each of the answers in your report:

- What was the survey or experimental unit?
- What were all the potential sources of variation?
- How was each of the individual sources compensated for? In particular, was the sample simple or stratified?
- How were subjects or units grouped into strata?
- Were the units sampled in clusters or individually?
- How were subjects assigned to experimental procedures? If assignment was at random, how was this accomplished?
- How was independence of the observations assured?

10.1.5   Clusters

Surveys often take advantage of the cost savings that result from naturally occurring groups, such as work sites, schools, clinics, neighborhoods, even entire towns or states. Not surprisingly, the observations within such a group are correlated. For example, individuals in the same community or work group often have shared views. Married couples, the ones whose marriages last, tend to have shared values. The effect of such correlation must be accounted for by the use of the appropriate statistical procedures. Thus, the nature and extent of such cluster sampling must be spelled out in detail in your reports.

> **Exercise 10.1:** Examine three recent reports in your field of study or in any field that interests you. (Examine three of your own reports if you have them.) Answer the following in each instance:
>
> 1. What was the primary hypothesis?
> 2. What was the principal alternative hypothesis?
> 3. Which statistics were employed and why?
> 4. Were all the assumptions underlying this statistic satisfied?
> 5. Was the power of this statistic reported?
> 6. Was the chosen statistic the most powerful available?
> 7. If the result was statistically significant, was the size of the effect estimated?
> 8. Was a one-tailed or two-tailed test used? Was this correct?
> 9. What was the total number of tests that were performed?

If the answers to these questions were not present in the reports you reviewed, what was the reason? Had their authors something to hide?

10.1.6   Validation Methods

A survey will be compromised if any of the following is true:

- Participants are not representative of the population of interest.
- Responses are not independent among respondents.
- Nonresponders, that is, those who decline to participate, would have responded differently.
- Respondents lie or answer carelessly.
- Forms are incomplete.

Your report should detail the preventive measures employed by the investigator and the degree to which they were successful.

You should describe the population(s) of interest in detail—providing demographic information where available, and similarly characterize the samples of participants to see whether they are indeed representative. (Graphs are essential here.)

You should describe the measures taken to ensure responses were independent including how participants were selected and where and how the survey took place.

A sample of nonrespondents should be contacted and evaluated. The demographics of the nonrespondents and their responses should be compared with those of the original sample.

How do you know respondents told the truth? You should report the results of any cross checks, such as redundant questions. And you should report the frequency of response omissions on a question-by-question basis.

10.2 TEXT, TABLE, OR GRAPH?

Whatever is the use of a book without pictures?

Alice in *Alice in Wonderland*

A series of major decisions need to be made as to how you will report your results—text, table, or graph? Whatever Alice's views on the subject, a graph may or may not be more efficient at communicating numeric information than the equivalent prose. This efficiency is in terms of the amount of information successfully communicated and not necessarily any space savings. Resist the temptation to enhance your prose with pictures.

And don't fail to provide a comprehensive caption for each figure. As Good and Hardin note in their Wiley text, *Common Errors in Statistics*, if the graphic is a summary of numeric information, then the graph caption is a summary of the graphic. The text should be considered part of the graphic design and should be carefully constructed rather than placed as an afterthought. Readers, for their own use, often copy graphics and tables that appear in articles and reports. A failure on your part to completely document the graphic in the caption can result in gross misrepresentation in these cases.

A sentence should be used for displaying two to five numbers, as in "The blood type of the population of the United States is approximately 45% O, 40% A, 11% B, and 4% AB." Note that the blood types are ordered by frequency.

Tables with appropriate marginal means are often the best method of presenting results. Consider adding a row (or column, or both) of contrasts; for example, if the table has only two rows, we could add a row of differences, row 1 minus row 2.

Tables dealing with two-factor arrays are straightforward, provided confidence limits are clearly associated with the correct set of figures. Tables involving three or more factors simultaneously are not always clear to the reader and are best avoided.

Make sure the results are expressed in appropriate units. For example, parts per thousand may be more natural than percentages in certain cases.

A table of deviations from row and column means as shown in Table 10.1 (or tables, if there are several strata) can alert us to the presence of outliers and may also reveal patterns in the data that were not yet considered.

Exercise 10.2: To report each of the following, should you use text, a table, or a graph? If a graphic, then what kind?

- Number of goals (each of 5 teams)
- Blood types of Australians
- Comparison treated/control red blood cell counts

Table 10.1   Residuals after Eliminating Main Effects

Sunlight		Fertilizer		Main effect
	Low	Medium	High	
Low				17
	4.1	−2.1	−11.2	
	9.1	4.1	−3.2	
	7.1	0.1	−7.2	
High	−9.8	−7.7	7.8	31.9
	−7.8	−0.7	12.8	
	−3.8	7.2	0.8	
Main Effect	8.3	25.3	39.7	

- Comparison of blood types in two populations
- Location leukemia cases by county
- Arm span versus height (six persons).

10.3 SUMMARIZING YOUR RESULTS

Your objective in summarizing your results should be to communicate some idea of all of the following:

1. The populations from which your samples are drawn.
2. Your estimates of population parameters.
3. The dispersion of the distribution from which the observations were drawn.
4. The precision of your estimates.

Proceed in three steps:

First, characterize the populations and subpopulations from which your observations are drawn. Of course, this is the main goal in studies of market segmentation. A histogram or scatterplot can help communicate the existence of such subpopulations to our readers. Few real-life distributions resemble the bell-shaped normal curve depicted in Figure 4.3. Most are bi- or even trimodal with each mode or peak corresponding to a distinct subpopulation. We can let the histogram speak for itself, but a better idea, particularly if you already suspect that the basis for market segments is the value of a second variable (such as home ownership, or level of education), is to add an additional dimension by dividing each of the histogram's bars into differently shaded segments whose size corresponds to the relative numbers in each subpopulation. Our R program to construct two overlaid histograms makes use of data frames, selection by variable value [hts$sex=="b",], and the options of the **hist()** function. Figure 10.1 is the result.

```
> height = c(141, 156.5, 162, 159, 157, 143.5, 154,
   158, 140, 142, 150, 148.5, 138.5, 161, 153, 145,
   147, 158.5, 160.5, 167.5, 155, 137)
```

Histogram of hts[hts$sex == "b",]$height

hts[hts$sex == "b",]$height

Figure 10.1   Histograms of class data by sex.

```
➢ sex = c("b",rep("g",7),"b",rep("g",6),rep("b",7))
➢ hts = data.frame(sex,height)
➢ hist(hts[hts$sex == "b",]$height,
  + xlim = c(min(height), max(height)), ylim = c(0,4),
  + col = "blue")
➢ hist(hts[hts$sex == "g",]$height,
  + xlim = c(min(height), max(height)), ylim = c(0,4),
  +add = T, border = "pink")
```

Similarly, we can provide for different subpopulations on a two-dimensional scatterplot by using different colors or shapes for the points.

```
➢ armspan = c(141, 156.5, 162, 159, 158, 143.5,
  155.5, 160, 140, 142.5, 148, 148.5, 139, 160,
  152.5, 142, 146.5, 159.5, 160.5, 164, 157, 137.5)
➢ psex = c(18,rep(25,7),18,rep(25,6),rep(18,7))
➢ plot(height,armspan, pch = psex)
```

yields Figure 10.2, while

```
➢ coplot(height~armspan|sexf)
➢ plot(sexf,height)
```

produces Figure 10.3.

Exercise 10.3: Redo Figure 10.2 and Figure 10.3 so as to provide legends and labels that will make it clear which set of data corresponds to each sex. Hint: Use R's **legend()** function.

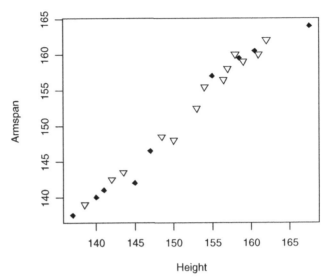

Figure 10.2 Overlying scatterplots of class data by sex.

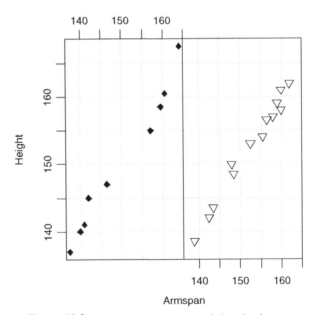

Figure 10.3 Side-by-side scatterplots of class data by sex.

10.3.1   Center of the Distribution

For small samples of three to five observations, summary statistics are virtually meaningless. Reproduce the actual observations; this is easier to do and more informative.

Though the arithmetic mean or average is in common use, it can be very misleading. For example, the mean income in most countries is far in excess of the *median* income or 50th percentile to which most of us can relate. When the arithmetic mean is meaningful, it is usually equal to or close to the median. Consider reporting the median in the first place.

The *geometric mean* is more appropriate than the arithmetic in three sets of circumstances:

1. When losses or gains can best be expressed as a percentage rather than a fixed value.
2. When rapid growth is involved.
3. When the data span several orders of magnitude, as with the concentration of pollutants.

Because bacterial populations can double in number in only a few hours, many government health regulations utilize the geometric rather than the arithmetic mean. A number of other government regulations also use it, though the sample median would be far more appropriate (and understandable). In any event, since your average reader may be unfamiliar with the geometric mean, be sure to comment in its use and on your reasons for adopting it.

The purpose of your inquiry must be kept in mind. Orders (in $) from a machinery plant ranked by size may be quite skewed with a few large orders. The median order size might be of interest in describing sales; the mean order size would be of interest in estimating revenues and profits.

Are the results expressed in appropriate units? For example, are parts per thousand more natural in a specific case than percentages? Have we rounded off to the correct degree of precision, taking account of what we know about the variability of the results, and considering whether the reader will use them, perhaps by multiplying by a constant factor, or another variable?

Whether you report a mean or a median, be sure to report only a sensible number of decimal places. Most statistical packages including R can give you nine or 10. Don't use them. If your observations were to the nearest integer, your report on the mean should include only a single decimal place. Limit tabulated values to no more than two effective (changing) digits. Readers can distinguish 354,691 and 354,634 at a glance but will be confused by 354,691 and 357,634.

10.3.2   Dispersion

The standard error of a summary is a useful measure of uncertainty *if* the observations come from a normal or Gaussian distribution (see Figure 4.3). Then, in 95%

of the samples, we would expect the sample mean to lie within two standard errors of the population mean.

But if the observations come from any of the following:

- a nonsymmetric distribution like an exponential or a Poisson
- a truncated distribution like the uniform
- a mixture of populations,

we *cannot* draw any such inference. For such a distribution, the probability that a future observation would lie between plus and minus one standard error of the mean might be anywhere from 40 to 100%.

Recall that the standard error of the mean equals the standard deviation of a single observation divided by the square root of the sample size. As the standard error depends on the squares of individual observations, it is particularly sensitive to outliers. A few excessively large observations, even a simple typographical error might have a dramatic impact on its value.

If you can't be sure your observations come from a normal distribution, then for samples from nonsymmetric distributions of size 6 or less, tabulate the minimum, the median, and the maximum. For samples of size 7 and up, consider using a box and whiskers plot. For samples of size 30 and up, the bootstrap may provide the answer you need.

10.3.3 Categorical Data

Always report actual numbers, not percentages. The differences shown in Table 10.2b are statistically significant; the differences shown in Table 10.2a are not; yet the relative frequencies are the same in both tables.

Exercise 10.4: Verify the correctness of the previous statement.

Table 10.2a   Effect of Treatment on Survival (Sample Size = 10, 10)

	Treat A	Treat B
Survived	5	9
Died	5	1

Table 10.2b   Effect of Treatment on Survival (Sample Size = 20, 20)

	Treat A	Treat B
Survived	10	18
Died	10	2

10.4 REPORTING ANALYSIS RESULTS

How you conduct and report your analysis will depend upon whether or not

- Baseline results of the various groups are equivalent.
- (if multiple observations sites were used) Results of the disparate experimental procedure sites may be combined.
- (if adjunct or secondary experimental procedures were used) Results of the various adjunct experimental procedure groups may be combined.
- Missing data, dropouts, and withdrawals are unrelated to experimental procedure.

Thus, your report will have to include

1. Demonstrations of similarities and differences for the following:
   - Baseline values of the various experimental procedure groups
   - Endpoints of the various subgroups determined by baseline variables and adjunct therapies.
2. Explanations of protocol deviations, including:
   - Ineligibles accidentally included in the study
   - Missing data
   - Dropouts and withdrawals
   - Modifications to procedures.

Further explanations and stratifications will be necessary if the rates of any of the above protocol deviations differ among the groups assigned to the various experimental procedures. For example, if there are differences in the baseline demographics, then subsequent results will need to be stratified accordingly. Moreover, some plausible explanation for the differences must be advanced.

Here is an example: Suppose the vast majority of women in the study were in the control group. To avoid drawing false conclusions about the men, the results for men and women must be presented separately, unless one first can demonstrate that the experimental procedures have similar effects on men and women.

Report the results for each primary endpoint separately. For each endpoint:

   a. Report the aggregate results by experimental procedure for all who were examined during the study for whom you have endpoint or intermediate data.
   b. Report the aggregate results by experimental procedure only for those subjects who were actually eligible, who were treated originally as randomized, or who were not excluded for any other reason. Provide significance levels for comparisons of experimental procedures.
   c. Break down these latter results into subsets based on factors determined prior to the start of the study as having potential impact on the response to treatment, such as adjunct therapy or gender. Provide significance

levels for comparisons of experimental procedures for these subsets of cases.

d. List all factors uncovered during the trials that appear to have altered the effects of the experimental procedures. Provide a tabular comparison by experimental procedure for these factors, but do *not* include *p*-values. The probability calculations that are used to generate *p*-values are not applicable to hypotheses and subgroups that are conceived *after* the data have been examined.

If there are multiple endpoints, you have the option of providing a further multivariate comparison of the experimental procedures.

Last, but by no means least, you must report the number of tests performed. When we perform multiple tests in a study, there may not be room (nor interest) to report all the results, but we do need to report the total number of statistical tests performed so that readers can draw their own conclusions as to the significance of the results that are reported. To repeat a finding of previous chapters, when we make 20 tests at the 1 in 20 or 5% significance level, we expect to find at least one or perhaps two results that are "statistically significant" by chance alone.

10.4.1   *p*-Values? Or Confidence Intervals?

As you read the literature of your chosen field, you will soon discover that *p*-values are more likely to be reported than confidence intervals. We don't agree with this practice, and here is why.

Before we perform a statistical test, we are concerned with its significance level, that is, the probability that we will mistakenly reject our hypothesis when it is actually true. In contrast to the significance level, the *p*-value is a random variable that varies from sample to sample. There may be highly significant differences between two populations and yet the samples taken from those populations and the resulting *p*-value may not reveal that difference. Consequently, it is not appropriate for us to compare the *p*-values from two distinct experiments, or from tests on two variables measured in the same experiment, and declare that one is more significant than the other.

If we agree in advance of examining the data that we will reject the hypothesis if the p-value is less than 5%, then our significance level is 5%. Whether our *p*-value proves to be 4.9 or 1 or 0.001%, we will come to the same conclusion. One set of results is not more significant than another; it is only that the difference we uncovered was measurably more extreme in one set of samples than in another.

We are less likely to mislead and more likely to communicate all the essential information if we provide confidence intervals about the estimated values. A confidence interval provides us with an estimate of the size of an effect as well as telling us whether an effect is significantly different from zero.

Confidence intervals, you will recall from Chapter 4, can be derived from the rejection regions of our hypothesis tests. Confidence intervals include all values of a parameter for which we would accept the hypothesis that the parameter takes that value.

Warning: A common error is to misinterpret the confidence interval as a statement about the unknown parameter. It is not true that the probability that a parameter is included in a 95% confidence interval is 95%. Nor is it at all reasonable to assume that the unknown parameter lies in the middle of the interval rather than toward one of the ends. What is true is that if we derive a large number of 95% confidence intervals, we can expect the true value of the parameter to be included in the computed intervals 95% of the time. Like the p-value, the upper and lower confidence limits of a particular confidence interval are random variables, for they depend upon the sample that is drawn.

The probability the confidence interval covers the true value of the parameter of interest and the method used to derive the interval must both be reported.

Exercise 10.5: Give at least two examples to illustrate why p-values are *not* applicable to hypotheses and subgroups that are conceived after the data are examined.

10.5 EXCEPTIONS ARE THE REAL STORY

Before you draw conclusions, be sure you have accounted for all missing data, interviewed nonresponders, and determined whether the data were missing at random or were specific to one or more subgroups.

Let's look at two examples, the first involving nonresponders, the second airplanes.

10.5.1 Nonresponders

A major source of frustration for researchers is when the variances of the various samples are unequal. Alarm bells sound. t-Tests and the analysis of variance are no longer applicable; we run to the advanced textbooks in search of some variance-leveling transformation. And completely ignore the phenomena we've just uncovered.

If individuals have been assigned at random to the various study groups, the existence of a significant difference in any parameter suggests that there is a difference in the groups. Our primary concern should be to understand why the variances are so different, and what the implications are for the subjects of the study. It may well be the case that a new experimental procedure is not appropriate because of higher variance, even if the difference in means is favorable. This issue is important whether or not the difference was anticipated.

In many clinical measurements, there are minimum and maximum values that are possible. If one of the experimental procedures is very effective, it will tend to push patient values into one of the extremes. This will produce a change in distribution from a relatively symmetric one to a skewed one, with a corresponding change in variance.

The distribution may not have a single mode but represents a mixture. A large variance may occur because an experimental procedure is effective for only a subset of the patients. When you compare mixtures of responders and nonresponders, specialized statistical techniques may be required.

10.5.2   The Missing Holes

During the Second World War, a group was studying planes returning from bombing Germany. They drew a rough diagram showing where the bullet holes were and recommended those areas be reinforced. Abraham Wald, a statistician, pointed out that essential data were missing. What about the planes that didn't return?

When we think along these lines, we see that the areas of the returning planes that had almost no apparent bullet holes have their own story to tell. Bullet holes in a plane are likely to be at random, occurring over the entire plane. The planes that did not return were those that were hit in the areas where the returning planes had no holes. Do the data missing from your own experiments and surveys also have a story to tell?

10.5.3   Missing Data

As noted in an earlier section of this chapter, you need to report the number and source of all missing data. But especially important is to summarize and describe all those instances in which the incidence of missing data varied among the various treatment and procedure groups.

The following are two examples where the missing data was the real finding of the research effort.

To increase participation, respondents to a recent survey were offered a choice of completing a printed form or responding online. An unexpected finding was that the proportion of missing answers from the online survey was half that from the printed forms.

A minor drop in cholesterol levels was recorded among the small fraction of participants who completed a recent trial of a cholesterol-lowering drug. As it turned out, almost all those who completed the trial were in the control group. The numerous dropouts from the treatment group had only unkind words for the test product's foul taste and undrinkable consistency.

10.5.4   Recognize and Report Biases

Very few studies can avoid bias at some point in sample selection, study conduct, and results interpretation. We focus on the wrong endpoints; participants and coinvestigators see through our blinding schemes; the effects of neglected and unobserved confounding factors overwhelm and outweigh the effects of our variables of interest. With careful and prolonged planning, we may reduce or eliminate many potential sources of bias, but seldom will we be able to eliminate all of them. Accept bias as inevitable and then endeavor to recognize and report all that do slip through the cracks.

Most biases occur during data collection, often as a result of taking observations from an unrepresentative subset of the population rather than from the population as a whole. An excellent example is the study that failed to include planes that did *not* return from combat.

When analyzing extended seismological and neurological data, investigators typically select specific cuts (a set of consecutive observations in time) for detailed analysis, rather than trying to examine all the data (a near impossibility). Not surprisingly, such "cuts" usually possess one or more intriguing features not to be found in run-of-the-mill samples. Too often theories evolve from these very biased selections.

The same is true of meteorological, geological, astronomical, and epidemiological studies, where with a large amount of available data, investigators naturally focus on the "interesting" patterns.

Limitations in the measuring instrument, such as censoring at either end of the scale, can result in biased estimates. Current methods of estimating cloud optical depth from satellite measurements produce biased results that depend strongly on satellite viewing geometry. Similar problems arise in high temperature and high-pressure physics and in radio-immune assay. In psychological and sociological studies, too often we measure that which is convenient to measure rather than that which is truly relevant.

Close collaboration between the statistician and the domain expert is essential if all sources of bias are to be detected, and, if not corrected, accounted for and reported. We read a report recently by economist Otmar Issing in which it was stated that the three principal sources of bias in the measurement of price indices are substitution bias, quality change bias, and new product bias. We've no idea what he was talking about, but we do know that we would never attempt an analysis of pricing data without first consulting an economist.

10.6 SUMMARY AND REVIEW

In this chapter, we discussed the necessary contents of your reports whether on your own work or that of others. We reviewed what to report, the best form in which to report it, and the appropriate statistics to use in summarizing your data and your analysis. We also discussed the need to report sources of missing data and potential biases. We provided some useful R functions for graphing data derived from several strata.

Chapter 11

Problem Solving

If you have made your way through the first 10 chapters of this text, then you may already have found that more and more people, strangers as well as friends, are seeking you out for your newly acquired expertise. (Not as many as if you were stunningly attractive or a film star, but a great many people nonetheless.) Your boss may even have announced that from now on you will be the official statistician of your group.

To prepare you for your new role in life, you will be asked in this chapter to work your way through a wide variety of problems that you may well encounter in practice. A final section will provide you with some over-all guidelines. You'll soon learn that deciding which statistic to use is only one of many decisions that need be made.

11.1 THE PROBLEMS

1. With your clinical sites all lined up and everyone ready to proceed with a trial of a new experimental vaccine versus a control, the manufacturer tells you that because of problems at the plant, the 10,000 ampoules of vaccine you've received are all he will be able to send you. Explain why you can no longer guarantee the power of the test.

2. After collecting some 50 observations, 25 on members of a control group, and 25 who have taken a low dose of a new experimental drug, you decide to add a third high-dose group to your clinical trial, and to take 75 additional observations, 25 on the members of each group. How would you go about analyzing these data?

3. You are given a data sample and asked to provide an interval estimate for the population variance. What two questions ought you to ask about the sample first?

4. John would like to do a survey of the use of controlled substances by teenagers but realizes he is unlikely to get truthful answers. He comes up with the following scheme: Each respondent is provided with a coin, instructions,

Introduction to Statistics Through Resampling Methods and R, Second Edition. Phillip I. Good.
© 2013 John Wiley & Sons, Inc. Published 2013 by John Wiley & Sons, Inc.

a question sheet containing two questions, and a sheet on which to write their answer, yes or no. The two questions are:

a. Is a cola (Coke or Pepsi) your favorite soft drink? Yes or no?

b. Have you used marijuana within the past seven days? Yes or no?

The teenaged respondents are instructed to flip the coin so that the interviewer cannot see it. If the coin comes up heads, they are to write their answer to the first question on the answer sheet; otherwise they are to write their answer to question 2.

Show that this approach will be successful, providing John already knows the proportion of teenagers who prefer colas to other types of soft drinks.

5. The town of San Philippe has asked you to provide confidence intervals for the recent census figures for their town. Are you able to do so? Could you do so if you had the some additional information? What might this information be? Just how would you go about calculating the confidence intervals?

6. The town of San Philippe has called on you once more. They have in hand the annual income figures for the past 6 years for their town and for their traditional rivals at Carfad-Sur-La-Mere and want you to make a statistical comparison. Are you able to do so? Could you do so if you had the some additional information? What might this information be? Just how would you go about calculating the confidence intervals?

7. You have just completed your analysis of a clinical trial and have found a few minor differences between patients subjected to the standard and revised procedures. The marketing manager has gone over your findings and noted that the differences are much greater if limited to patients who passed their first postprocedure day without complications. She asks you for a p-value. What do you reply?

8. At the time of his death in 1971, psychologist Cyril Burt was viewed as an esteemed and influential member of his profession. Within months, psychologist Leon Kamin reported numerous flaws in Burt's research involving monozygotic twins who were reared apart. Shortly thereafter, a third psychologist, Arthur Jensen also found fault with Burt's data.

Their primary concern was the suspicious consistency of the correlation coefficients for the intelligence test scores of the monozygotic twins in Burt's studies. In each study, Burt reported sum totals for the twins he had studied so far. His original results were published in 1943. In 1955, he added six pairs of twins and reported results for a total of 21 sets of twins. Likewise, in 1966, he reported the results for a total of 53 pairs. In each study, Burt reported correlation coefficients indicating the similarity of intelligence scores for monozygotic twins who were reared apart. A high correlation coefficient would make a strong case for Burt's hereditarian views.

Burt reported the following coefficients: 1943: $r = 0.770$; 1955: $r = 0.771$; 1966: $r = 0.771$. Why was this suspicious?

9. Which hypothesis testing method would you use to address each of the following? Permutation, parametric, or bootstrap?
   a. Testing for an ordered dose response.
   b. Testing whether the mean time to failure of new light bulb in intermittent operation is one year.
   c. Comparing two drugs using the data from the following contingency table.

	Drug A	Drug B
Respond	5	9
No	5	1

   d. Comparing old and new procedures using the data from the following 2 × 2 factorial design

	Control	Old
Control		1150
		2520
		900
		50
Young	5640	7100
	5120	11,020
	780	13,065
	4430	
	7230	

Ethical Standard

Polish-born Jerzy Neyman (1894–1981) is generally viewed as one of the most distinguished statisticians of the twentieth century. Along with Egon Pearson, he is responsible for the method of assigning the outcomes of a set of observations to either an acceptance or a rejection region in such a way that the power is maximized against a given alternative at a specified significance level. He was asked by the U.S. government to be part of an international committee monitoring the elections held in a newly liberated Greece after World War II. In the oversimplified view of the U.S. State Department, there were two groups running in the election: The Communists and The Good Guys. Professor Neyman's report that both sides were guilty of extensive fraud pleased no one but set an ethical standard for other statisticians to follow.

---

10. The government has just audited 200 of your company's submissions over a 4-year period and has found that the average claim was in error in the amount of $135. Multiplying $135 by the 4000 total submissions during that period, they are asking your company to reimburse the government in

the amount of $540,000. List all possible objections to the government's approach.

11. Since I first began serving as a statistical consultant almost 50 years ago, I've made it a practice to begin every analysis by first computing the minimum and maximum of each variable. Can you tell why this practice would be of value to you as well?

12. Your mother has brought your attention to a newspaper article in which it is noted that one school has successfully predicted the outcome of every election of a U.S. president since 1976. Explain to her why this news does not surprise you.

13. A clinical study is well under way when it is noted that the values of critical end points vary far more from subject to subject than was expected originally. It is decided to increase the sample size. Is this an acceptable practice?

14. A clinical study is well under way when an unusual number of side effects is observed. The treatment code is broken and it is discovered that the majority of the effects are occurring to subjects in the control group. Two cases arise:
   a. The difference between the two treatment groups is statistically significant. It is decided to terminate the trials and recommend adoption of the new treatment. Is this an acceptable practice?
   b. The difference between the two treatment groups is *not* statistically significant. It is decided to continue the trials but to assign twice as many subjects to the new treatment as are placed in the control group. Is this an acceptable practice?

15. A jurist has asked for your assistance with a case involving possible racial discrimination. Apparently the passing rate of minorities was 90% compared with 97% for whites. The jurist didn't think this was much of a difference, but then one of the attorneys pointed out that these numbers represented a jump in the failure rate from 3% to 10%. How would you go about helping this jurist to reach a decision?

   When you hired on as a statistician at the Bumbling Pharmaceutical Company, they told you they'd been waiting a long time to find a candidate like you. Apparently they had, for your desk is already piled high with studies that are long overdue for analysis. Here is just a sample:

16. The endpoint values recorded by one physician are easily 10 times those recorded by all other investigators. Trying to track down the discrepancies, you discover that this physician has retired and closed his office. No one knows what became of his records. Your coworkers instantly begin to offer you advice including all of the following:
   a. Discard all the data from this physician.
   b. Assume this physician left out a decimal point and use the corrected values.

c. Report the results for this observer separately.

d. Crack the treatment code and then decide.

What will you do?

17. A different clinical study involved this same physician. This time, he completed the question about side effects that asked, is this effect "mild, severe, or life threatening," but failed to answer the preceding question that specified the nature of the side effect. Which of the following should you do?

    e. Discard all the data from this physician.

    f. Discard all the side effect data from this physician.

    g. Report the results for this physician separately from the other results.

    h. Crack the treatment code and then decide.

18. Summarizing recent observations on the planetary systems of stars, the Monthly Notices of the Royal Astronomical Society reported that the vast majority of extra-solar planets in our galaxy must be gas giants like Jupiter and Saturn as no Earth-size planet has been observed. What is your opinion?

11.2 SOLVING PRACTICAL PROBLEMS

In what follows, we suppose that you have been given a data set to analyze. The data did not come from a research effort that you designed, so there may be problems, many of them. We suggest you proceed as follows:

1. Determine the provenance of the observations.

2. Inspect the data.

3. Validate the data collection methods.

4. Formulate your hypotheses in testable form.

5. Choose methods for testing and estimation.

6. Be aware of what you don't know.

7. Perform the analysis.

8. Qualify your conclusions.

11.2.1 Provenance of the Data

Your very first questions should deal with *how* the data were collected. What population(s) were they drawn from? Were the members of the sample(s) selected at random? Were the observations independent of one another? If treatments were involved, were individuals assigned to these treatments at random? Remember, statistics is applicable only to random samples.* You need to find out all the details of the sampling procedure to be sure.

---

* The one notable exception is that it is possible to make a comparison between entire populations by permutation means.

You also need to ascertain that the sample is representative of the population it purports to be drawn from. If not, you'll need either to (1) weight the observations, (2) stratify the sample to make it more representative, or (3) redefine the population before drawing conclusions from the sample.

11.2.2   Inspect the Data

If satisfied with the data's provenance, you can now begin to inspect the data you've been provided. Your first step should be to compute the minimum and the maximum of each variable in the data set and to compare them with the data ranges you were provided by the client. If any lie outside the acceptable range, you need to determine which specific data items are responsible and have these inspected and, if possible, corrected by the person(s) responsible for their collection.

I once had a long-term client who would not let me look at the data. Instead, he would merely ask me what statistical procedure to use next. I ought to have complained, but this client paid particularly high fees, or at least he did so in theory. The deal was that I would get my money when the firm for which my client worked got its first financing from the venture capitalists. So my thoughts were on the money to come and not on the data.

My client took ill—later I was to learn he had checked into a rehabilitation clinic for a metamphetamine addiction—and his firm asked me to take over. My first act was to ask for my money—they'd got their financing. While I waited for my check, I got to work, beginning my analysis as always by computing the minimum and the maximum of each variable. Many of the minimums were zero. I went to verify this finding with one of the technicians only to discover that zeros were well outside the acceptable range.

The next step was to look at the individual items in the database. There were zeros everywhere. In fact, it looked as if more than half the data were either zeros or repeats of previous entries. Before I could report these discrepancies to my client's boss, he called me in to discuss my fees.

"Ridiculous," he said. We did not part as friends. I almost regret not taking the time to tell him that half the data he was relying on did not exist. Tant pis. No, this firm is not still in business.

Not incidentally, the best cure for bad data is prevention. I strongly urge that all your data be entered directly into a computer so it can be checked and verified immediately upon entry. You don't want to be spending time tracking down corrections long after whoever entered 19.1 can remember whether the entry was supposed to be 1.91 or 9.1 or even 0.191.

11.2.3   Validate the Data Collection Methods

Few studies proceed exactly according to the protocol. Physicians switch treatments before the trial is completed. Sets of observations are missing or incomplete. A measuring instrument may have broken down midway through and been replaced by another, slightly different unit. Scoring methods were modified and observers provided with differing criteria employed. You need to determine the ways in which the protocol was modified and the extent and impact of such modifications.

A number of preventive measures may have been employed. For example, a survey may have included redundant questions as cross checks. You need to determine the extent to which these preventive measures were successful. Was blinding effective? Or did observers crack the treatment code? You need to determine the extent of missing data and whether this was the same for all groups in the study. You may need to ask for additional data derived from follow-up studies of nonresponders and dropouts.

11.2.4 Formulate Hypotheses

All hypotheses must be formulated *before* the data are examined. It is all too easy for the human mind to discern patterns in what is actually a sequence of totally random events—think of the faces and animals that seem to form in clouds.

As another example, suppose that while just passing the time you deal out a five-card poker hand. It's a full house! Immediately, someone exclaims "what's the probability that could happen?" If by "that" a full house is meant, its probability is easily computed. But the same exclamation might have resulted had a flush or a straight been dealt or even three of a kind. The probability that "an interesting hand" will be dealt is much greater than the probability of a full house. Moreover, this might have been the third or even the fourth poker hand you've dealt; it's just that this one was the first to prove interesting enough to attract attention.

The details of translating objectives into testable hypotheses were given in Chapter 6 and Chapter 9.

11.2.5 Choosing a Statistical Methodology

For the two-sample comparison, a *t*-test should be used. Remember, one-sided hypotheses lead to one-sided tests and two-sided hypotheses to two-sided tests. If the observations were made in pairs, the paired *t*-test should be used.

Permutation methods should be used to make *k*-sample and multivariate comparisons. Your choice of statistic will depend upon the alternative hypothesis and the loss function.

Permutation methods should be used to analyze contingency tables.

The bootstrap is of value in obtaining confidence limits for quantiles and in model validation.

When multiple tests are used, the *p*-values for each test are no longer correct but must be modified to reflect the total number of tests that were employed. For more on this topic, see P. I. Good and J. Hardin, *Common Errors in Statistics*, 4th ed. (New York: Wiley, 2012).

11.2.6   Be Aware of What You Don't Know

Far more statistical theory exists than can be provided in the confines of an introductory text. Entire books have been written on the topics of survey design, sequential analysis, and survival analysis, and that's just the letter "s." If you are unsure what statistical method is appropriate, don't hesitate to look it up on the Web or in a more advanced text.

11.2.7   Qualify Your Conclusions

Your conclusions can only be applicable to the extent that samples were representative of populations and experiments and surveys were free from bias. A report by GC Bent and SA Archfield is ideal in this regard.* This report can be viewed online at http://water.usgs.gov/pubs/wri/wri024043/.

They devote multiple paragraphs to describing the methods used, the assumptions made, the limitations on their model's range of application, potential sources of bias, and the method of validation. For example, "The logistic regression equation developed is applicable for stream sites with drainage areas between 0.02 and $7.00\,mi^2$ in the South Coastal Basin and between 0.14 and $8.94\,mi^2$ in the remainder of Massachusetts, because these were the smallest and largest drainage areas used in equation development for their respective areas.

The equation may not be reliable for losing reaches of streams, such as for streams that flow off area underlain by till or bedrock onto an area underlain by stratified-drift deposits . . .

The logistic regression equation may not be reliable in areas of Massachusetts where ground-water and surface-water drainage areas for a stream site differ." (Brent and Archfield provide examples of such areas.)

This report also illustrates how data quality, selection, and measurement bias can affect results. For example,

The accuracy of the logistic regression equation is a function of the quality of the data used in its development. This data includes the measured perennial or intermittent status of a stream site, the occurrence of unknown regulation above a site, and the measured basin characteristics.

The measured perennial or intermittent status of stream sites in Massachusetts is based on information in the USGS NWIS database. Stream- flow measured as less than 0.005 ft3/s is rounded down to zero, so it is possible that several streamflow measurements reported as zero may have had flows less than 0.005 ft3/s in the stream. This measurement would cause stream sites to be classified as intermittent when they actually are perennial.

It is essential that your reports be similarly detailed and qualified whether they are to a client or to the general public in the form of a journal article.

---

* A logistic regression equation for estimating the probability of a stream flowing perennially in Massachusetts USGC. Water-Resources Investigations Report 02-4043.

Answers to Selected Exercises

Exercise 2.11: One should always test R functions one has written by trying several combinations of inputs. Will your function work correctly if $m > n$? $m < 0$? Be sure to include solution messages such as

```
> print("m must be less than or equal to n"); break
```

Exercise 2.13:

$$P(T|R) = 1 - P(T^c|R).$$

Exercise 3.7:

$$\sum_{i=1}^{n}(X_i - \bar{X}_.) = \sum_{i=1}^{n} X_i - \sum_{i=1}^{n} \bar{X}_{.i} = \sum_{i=1}^{n} X_i - n\bar{X}_.$$

Exercise 3.14:

$$
\begin{aligned}
\sum_{k=0}(k-\lambda)^2 \Pr\{x=k\} &= \sum_{k=0}(k^2 - 2k\lambda + \lambda^2)^2 \Pr\{x=k\} \\
&= \sum_{k=0} k^2 \Pr\{x=k\} - 2\lambda\sum_{k=0} k \Pr\{x=k\} + \lambda^2 \sum_{k=0}\Pr\{x=k\} \\
&= \sum_{k=1} k^2\lambda^k e^{-\lambda} / k! - \lambda^2 \\
&= \sum_{k=1} k\lambda\lambda^{k-1} e^{-\lambda} / (k-1)! - \lambda^2 \\
&= \lambda\sum_{k=1}[1 + (k-1)]\lambda^{k-1} e^{-\lambda} / (k-1)! - \lambda^2 \\
&= \lambda\left\{\sum_{k=1}\lambda^{k-1} e^{-\lambda} / (k-1)! + \sum_{k=1}(k-1)\lambda^{k-1} e^{-\lambda} / (k-1)!\right\} - \lambda^2 \\
&= \lambda\sum_{k=0}\lambda^k e^{-\lambda} / k! + \lambda^2 \sum_{k=2}\lambda^{k-2} e^{-\lambda} / (k-2)! - \lambda^2 \\
&= \lambda + \lambda^2 - \lambda^2
\end{aligned}
$$

Exercise 4.7:

```
sales=c(754,708,652,783,682,778,665,799,693,825,828,
674,792,723)
t.test(sales, conf.level=0.9)
```

Introduction to Statistics Through Resampling Methods and R, Second Edition. Phillip I. Good.
© 2013 John Wiley & Sons, Inc. Published 2013 by John Wiley & Sons, Inc.

Exercise 5.4: The results will have a binomial distribution with 20 trials; p will equal 0.6 under the primary hypothesis, while under the alternative p will equal 0.45.

Exercise 5.5: You may want to reread Chapter 2 if you had trouble with this exercise.

$$\text{Pr\{at least one Type I errors\}} = 1 \quad \text{Pr\{no type I errors\}} = 1 - .95^{20}.$$

Exercise 8.2:

```
> imp=c(2, 4, 0.5, 1, 5.7, 7, 1.5, 2.2)
> log(imp)
[1]  0.6931472  1.3862944
-0.6931472  0.0000000  1.7404662  1.9459101  0.4054651
[8]  0.7884574
> t.test(log(imp))
      One Sample t-test
data: log(imp)
t = 2.4825, df = 7, p-value = 0.04206
alternative hypothesis: true mean is not equal to 0
95 percent confidence interval: 0.03718173 1.52946655
sample estimates: mean of x 0.7833241
#Express this mean as an after/before ratio
> exp(0.7833241)
[1]  2.188736
```

Exercise 8.11:

```
#Define F1
> Snoopy = c(2.5, 7.3, 1.1, 2.7, 5.5, 4.3)
> Quick = c(2.5, 1.8, 3.6, 4.2 , 1.2 ,0.7)
> Mrs. Good's = c(3.3, 1.5, 0.4, 2.8, 2.2 ,1.0)
> size = c(length(Snoopy), length(Quick),
length(MrsGood))
> data = c(Snoopy, Quick, MrsGood)
> f1=F1(size,data)
> N = 1600
> cnt = 0
> for (i in 1:N){
+       pdata = sample (data)
+       f1p=F1(size,pdata)
+       # counting number of rearrangements for which F1
greater than or equal to original
+       if (f1 <= f1p) cnt=cnt+1
+       }
> cnt/N
[1]  0.108125
```

Exercise 9.17: This example illustrates the necessity of establishing a causal relationship between the predictor and the variable to be predicted before attempting to model the relationship.

Index

Introduction to Statistics Through Resampling Methods and R, Second Edition. Phillip I. Good.
© 2013 John Wiley & Sons, Inc. Published 2013 by John Wiley & Sons, Inc.

Printed and bound by CPI Group (UK) Ltd, Croydon, CR0 4YY

27/10/2024